Time in Ecology

MONOGRAPHS IN POPULATION BIOLOGY

SIMON A. LEVIN AND HENRY S. HORN, SERIES EDITORS

Time in Ecology
A Theoretical Framework

Eric Post

PRINCETON UNIVERSITY PRESS

Princeton and Oxford

Copyright © 2019 by Princeton University Press

Published by Princeton University Press
41 William Street, Princeton, New Jersey 08540
6 Oxford Street, Woodstock, Oxfordshire OX20 1TR

press.princeton.edu

All Rights Reserved

Library of Congress Control Number 2018931071
Cloth ISBN 978-0-691-16386-4
Paperback ISBN 978-0-691-18235-3

British Library Cataloging-in-Publication Data is available

This book has been composed in Times Roman

Printed on acid-free paper. ∞

Printed in the United States of America

10 9 8 7 6 5 4 3 2 1

For Boochie, Phoebe, and Mason. According to some

theories, time for them lies mostly in the future.

Tempus rerum imperator

Physics, and physics alone, is the science that actually considers time itself to be a target of study.
—Callender, *The Oxford Handbook of Philosophy of Time*

Contents

Acknowledgments   xi

Introduction. A Framework for the Role of Time in Ecology   1

1. What Is Time?   7
   Philosophical Views of Time: Idealism, Relationism, and Realism   7
   A-Series, B-Series, and the Flow of Time   9
   Space-Time and Astrophysical Theories of Time   11

2. Phenological Advance, Stasis, and Delay   15
   Phenology versus Seasonality   15
   Ubiquitous Variability in Phenological Trends   20
   Phenological Dynamics across Taxa, Latitude, and Time   26
   Variation in Rates of Land-surface Warming across Latitude and Time   37

3. Ecological Time   43
   Time as a Resource in Ecology   45
   Scales of Time in the Domain of Phenology   50
   Timing and Duration of Life History Events   53
   The Adaptive Value of Phenological Advance, Stasis, and Delay   55
   Parallels and Contrasts between the Notions of Space and Time
     as Resources   59
   Three Forms of Time: Cosmological, Recurrent, and Relative
     Ecological Time   60
   The Space-Time Continuum in a Life History Context   63
   Fragmentation of Time   64

4. The Phenological Niche   67
   The Absolute Phenological Niche   67
   The Relative Phenological Niche   71
   Mutualisms and the Relative Phenological Niche   82
   Biodiversity Implications of Phenological Niche Conservatism
     in Specialist Associations   85
   Biodiversity Implications of Phenological Niche Conservatism
     in Generalist Associations   89

5. The Phenological Community   94
        The Compositional versus Phenological Community   94
        Phenological Community Dynamics and Species' Rank Order   98
        Niche Packing and Changes in the Availability of Ecological Time 104

6. Use of Time in the Phenology of Horizontal Species Interactions   107
        Timing and Duration: Responses to Potentially Distinct
            Selection Pressures   107
        Variation in Timing within and among Individuals and Species   112
        Variation in Timing among Species within Communities   117
        Empirical Applications to Horizontal Phenological Communities   119
        Indices of Phenological Overlap among Species   124

7. Use of Time in the Phenology of Vertical Species Interactions   133
        The Use of Time in Bitrophic-level Systems   134
        Applications to Mutualisms   138
        The Use of Time in Tritrophic-level Systems   139
        Phenological Cascades   142
        Empirical Assessment of Phenological Dynamics
            in Bitrophic-level Systems   143
        An Empirical Example of a Phenological Cascade   155
        Additional Examples of Changes in the Use of Time
            in Tritrophic-level Systems   159
        Final Considerations of the Use of Time
            in Pollinator-Plant Associations   163

8. Limitations and Extension to Tropical Systems   167
        Seasonal and Interannual Phenological Dynamics in Tropical Systems 170
        Partitioning of Time in Phenological Communities in the Tropics   176
        Availability of Time Related to Energy Input   178

9. The More General Role of Time in Ecology   180

Appendix A. Online Resources of Relevance to Phenology   187

Appendix B. Sources Used in the Meta-analysis in Chapter 2   191

References   195

Index   221

Acknowledgments

This book is about a proposed relationship between time and phenology. I owe my interest in these subjects, and my conviction that the relationship between them warrants a formalized treatment, to the influences of several people over the past three decades.

As a college freshman at the University of Wisconsin—Stevens Point, I enrolled in a course in introductory astronomy that led me to wonder about the nature of time. The professor who taught that course, Mark Bernstein, patiently answered many questions about time and relativity during his office hours that semester. After transferring to the University of Minnesota the following year, I lost touch with Mark, but his lessons and our conversations have lingered on in my memory as the years have passed. In hindsight, I can see that those interactions were the genesis of many of the thoughts that spurred the ideas in this book.

Early in graduate school, it wouldn't have occurred to me to study phenology if not for the influences of my advisor, Dave Klein, and Terry Bowyer. Dave encouraged me to incorporate some consideration of plant phenology into my dissertation research on caribou foraging ecology. Initially, I had difficulty imagining the role plant phenology might play in the ecology and demography of caribou because I naïvely considered vegetation to be a largely time-invariant resource over the lifetime of an individual herbivore. Terry Bowyer opened my eyes to the importance of plant phenology to the timing of reproduction and reproductive success of herbivores in seasonal environments. Terry convinced me that a project linking caribou parturition phenology and plant phenology would be an interesting and worthwhile pursuit.

Since then, many individuals have assisted tirelessly in the collection of phenological data on plants, caribou, and muskoxen at my study site near Kangerlussuaq, Greenland. There are probably few things in ecology more monotonous than collecting such data, so I'm grateful and indebted to all of them. Of special importance among these are Pernille Sporon Bøving, our son, Mason, and our daughter, Phoebe. Anyone who leaves family behind to go into the field, or who is left behind by a loved one doing fieldwork, knows the sacrifice this entails. I've been fortunate to have the comfort of close family alongside me in the field for most of my career. Additional friends, students, and colleagues who have maintained the steady flow of phenology data and insights at study sites in Alaska and

Greenland over the years include Mike Avery, Emma Behr, Eva Beyen, Jesper Bahrenscheer, Sean Cahoon, Megan Eberbach, Mads Forchhammer, Nell Herrmann, Conor Higgins, Didem Ikis, Toke Høye, Christian John, Syrena Johnson, Jeff Kerby, Janine Mistrick, Frank Mörschel, Christian Pedersen, Ieva Perkons, Taylor Rees, Henning Thing, David Watts, Chris Wilmers, and Tyler Yenter.

Several colleagues, whether directly through prolonged interactions or indirectly through the literature, have had prominent, guiding influences on my thinking about phenology, and how to study it, and to them I am indebted. These include Julio Betancourt, Christiaan Both, Paul CaraDonna, Mads Forchhammer, Toke Høye, David Inouye, Christian John, Mark Hebblewhite, Amy Iler, Jeff Kerby, Camille Parmesan, Richard Primack, Andrew Richardson, Terry Root, Abe Miller-Rushing, Mark Schwartz, Heidi Steltzer, Nils Christian Stenseth, Stephen Thackeray, Henning Thing, Marcel Visser, David Watts, and Lizzie Wolkovich. I am also grateful to David Watts and Nick Tyler for providing blunt and constructive feedback on some of the core ideas in this book, as those ideas were still in the early stages of development.

For contributing data and reprints of publications that were difficult to locate and essential to some of the content and insights in this book, heartfelt thanks go to Ignacio Bartomeus, Laura Burkle, Ben Cook, Charles Davis, Dave Mackas, Richard Primack, Lizzie Wolkovich, and Xiaoyang Zhang. Lizzie Wolkovich and Elsa Cleland also generously contributed figure 4.1. I am grateful to Byron Steinman and Michael Mann for their guidance, feedback, insights, and assistance with analyses that were of critical importance to the meta-analysis reviewed in chapter 2.

I would also like to thank an anonymous reader of two previous drafts of this book for highlighting gaps in my reasoning, and for encouraging me to think more about extensions of the theoretical framework developed in this book to tropical systems. Likewise, I'm grateful for having had the opportunity to present some of the ideas in this book while they were still in development at departmental seminars at the University of Nevada, Reno; the University of California, Davis; and the University of California, Santa Cruz. Colleagues at each of these institutions were generously forthcoming with critical feedback as well as encouragement. In particular, I would like to thank Lee Dyer, Paul Hurtado, and Nick Pardikes at UNR; John Eadie, Susan Harrison, Marcel Holyoak, Art Shapiro, Andy Sih, Mark Schwartz, and Sharon Strauss at UC Davis; and Jim Estes, Jeff Kiehl, Michael Loik, and Chris Wilmers at UC Santa Cruz for thought-provoking comments and questions. I am also grateful to Alison Kalett at Princeton University Press for her patience with the revision process. Alison's encouragement, support, and interest in this project were essential to its development and timely completion. Henry Horn also offered valuable guidance on improving the presentation of the book's core concepts. Alison, Henry, Jim Estes, and Chris Wilmers all offered insightful suggestions for the book's title. Last, thanks to Jennifer Harris for meticulous copyediting of the final manuscript.

Phenology data are most informative when they are continuous and long term, yet funding agencies seem reluctant to fund long-term projects. I am grateful for continuous funding of my research on phenology through multiple short-term grants since 1991 from the University of Alaska Fairbanks Chancellor's Graduate Fellowship and Graduate Natural Resources Fellowship, the U.S. National Science Foundation (NSF), the Norwegian Science Council, the U.S. Fish & Wildlife Service, the Committee for Research and Exploration at National Geographic, and the National Aeronautics and Space Administration (NASA).

Time in Ecology

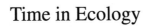

A Framework for the Role
of Time in Ecology

Imagine a dimensionless universe, one devoid of time and space. Now try to imagine the nature of ecology and evolution in such a universe. Would there be pattern, process, or dynamics of any kind? Unlikely, to say the least. If we add a spatial dimension to such a universe, but leave it static with respect to time, would this make a difference? In the absence of any existing variation to superimpose upon the spatial dimension, how would variation across space arise without time? In contrast, if we began with a dimensionless universe but then added time to it, variation, and in turn pattern, process, and dynamics may very well develop even in the absence of space. But how does time explain ecological pattern and process? This book is intended to develop a framework for a novel way of thinking about time in ecology, using the study of phenology as an exemplar for doing so. Hence, the book may also encourage novel ways of thinking about phenology and about life history dynamics in general.

Since the publication of *Ecology of Climate Change* in 2013, the Earth has continued to warm, and in fact, during the writing of the first draft of this book, experienced its warmest year on record, in 2015 (NOAA 2016). However, that warming record itself was surpassed in 2016 (Potter et al. 2017). Not coincidentally, there have also appeared in the literature recent multiple compelling accounts of the continued progression among an array of species and biomes toward earlier and earlier onset of springtime events. Perhaps most notable among these is the report, in 2013, of record-early flowering by 27 of 32 species of early-spring flowering plants in Massachusetts, and by 19 of 23 early-spring flowering plants in Wisconsin (Ellwood et al. 2013). The oldest records of flowering times for these species at those sites date back to observations by Henry David Thoreau in Massachusetts between 1852 and 1858, and by Aldo Leopold in Wisconsin between 1934 and 1945; spring temperatures since then have warmed at those locations by 2.5°C and 1.8°C, respectively (Ellwood et al. 2013).

But advances in the timing of biological events in association with warming are neither universal across taxa nor within taxa across sites. There also appears to be a growing emphasis in studies of phenology, the discipline that concerns itself with the timing of events, on patterns of delayed autumn phenology and the role of this in lengthening growing seasons in the northern hemisphere (Jeong et al. 2011; Archetti et al. 2013; Richardson et al. 2013; Garonna et al. 2014; Tang et al. 2015). For instance, a recent analysis of global trends in plant phenological dynamics utilizing three decades of satellite-derived Normalized Difference Vegetation Index (NDVI) data concluded that trends in end of season phenology were generally stronger than those in start of season phenology, and contributed relatively more to trends in annual growing season length (Garonna et al. 2016). Additional examples of delayed onset or occurrence of phenological events associated directly or indirectly with warming have been documented in butterflies (Altermatt 2012; Diamond et al. 2014; Karlsson 2014), birds (Beaumont et al. 2006; Lee et al. 2011), plants (Prieto et al. 2009; Bokhorst et al. 2011; Liancourt et al. 2012; Dorji et al. 2013; Ishioka et al. 2013; Laube et al. 2014; Bjorkman et al. 2015; Marchin et al. 2015; Rawal et al. 2015; Mulder et al. 2017), dragonflies (Doi 2008), grasshoppers (Forrest 2016), penguins (Hindell et al. 2012), noctuid moths (Liu et al. 2011), intertidal gastropods (Moore et al. 2011), and leatherback turtles (Neeman et al. 2015), to name a few. Additionally, recent analyses of satellite NDVI data indicate that phenological dynamics across over half of the Earth's land surface have changed by more than two standard deviations since 1981 (Buitenwerf et al. 2015).

Such patterns complement an existing body of work in this field that has also emphasized the absence of any discernable phenological trends in some traits, species, or study locales (Hart et al. 2014). For instance, a long-term observational study of first spring flight dates of 23 species of butterflies in California reported a mean advance of 24 days over 31 years among the four species undergoing significant advances in flight dates (Forister and Shapiro 2003). The same study reported, however, no significant change in first spring flight dates in the remaining 19 species (Forister and Shapiro 2003). Furthermore, an updated analysis of an extension of this data set that included observations through 2015 reported significant delays in first spring flight dates in two of the species monitored (Forister and Shapiro in press). Hence, although phenological advance appears more commonly in the literature, delays and stasis are not entirely uncommon.

Phenology has long been studied in the context of dynamical responses to the alleviation of environmental constraints on the expression of life history traits related to timing (Sørensen 1941; Caprio 1957; Lack 1966; Goff and Cole 1976; Harris 1977; Sugg et al. 1983; Breeman et al. 1988; Stamou et al. 1993; Silvertown et al. 1997; Adler et al. 2014). Most commonly, such constraints embody limits on the timing of biological activity imposed by photoperiod at high latitudes, or solar irradiance at lower latitudes; temperature; moisture or precipitation; or some combination of these. This book diverges somewhat from this well-established and long-standing view of phenology as a response dynamic. It will encourage a complementary view of phenology as the expression of an active strategy aimed at capturing and allocating an overlooked resource: time. The potentially controversial notion that time is a biological resource in and of itself is critical to making sense of the fact that, while phenological advances are widespread across taxa and biomes, they are not universal. This notion should help us understand why, in response to the same environmental stimulus, a diversity of phenological responses may ensue, and why this diversity is evident at the organismal level, the species level, and the community level. For instance, whether in response to drought, in response to warming, in response to variation in cloudiness and solar irradiance, or in response to snow melt timing, the timing of some life history events within an individual may advance while others become delayed or remain fixed. Similarly, the same sorts of environmental changes may elicit variable rates of advance or delay, or no response at all, across individuals within populations of a species. Or they may elicit different phenological responses among species at the same site. How do we explain such variability in an ecological and evolutionary context? Traditionally, we may view such patterns as adaptive phenological plasticity in response to variation in environmental seasonality. But we may also recognize such patterns as adaptive *strategies* once we view time as a resource, the allocation of which to development, maintenance, production, and reproduction determines fitness.

Ecology has circled around and brushed up against this notion for decades. It began with the idea that time is just one of many axes in the n-dimensional hypervolume of the niche along which species may segregate to minimize competition for other resources (Schoener 1974). It surfaced soon thereafter in a treatment of butterfly phenology that observed that, in contrast to patterns seen in insects, the activity patterns of mammals and birds "are so nearly synchronous that time can almost be ruled out as a resource to be sub-divided among them" (Shapiro 1975). And it has subsequently progressed through discussions of the "meaning of time" in metabolic rates and life spans of individuals (Schmidt-Nielsen 1984), metabolic scaling laws (West et al. 1997; Brown et al. 2004), and partitioning of time by interacting organisms (Kronfeld-Schor and Dayan 2003). More recently, ecologists have encouraged the development of frameworks for the treatment of time as one of the two major axes defining and determining ecological dynamics and patterns (the other being space) (Kelly et al. 2013; Wolkovich et al. 2014b).

These arguments can be refined and nudged further toward a view of time that brings into clearer focus its functional role in ecology and evolution. This view has the potential to transform our conceptualization of and perception of time in ecology from that of a simple measure of occurrence and rate to that of a major driver of the evolution of life history strategies and their variable expression. In essence, this transformation requires the development of a convincing argument for the case that time is not only a resource but also that time may in fact be the only resource of truly limited availability. This latter point rests on the notion that time, unlike other resources, cannot be stored. It can be used only to allocate or convert energy to other forms. Plants, for instance, make a living by converting time, solar energy, and carbon dioxide to biomass and offspring.

During the development of the ensuing theoretical framework, I have tried to be comprehensive in my thinking about arguments against this line of reasoning. The most obvious of these is that time is actually a construct of human consciousness and, as such, may not in actuality exist independently of human awareness (McTaggart 1908; Schultze 1908; Robertson 1923). If time does not in fact exist, such an argument might go, then it cannot possibly be of use to living things, much less represent a resource. The cosmological theory of time, which argues that the apparent forward progression of time is a consequence of the expansion of the universe (Hawking 1969, 1985), suggests, however, that time does exist independently of human awareness. More practical arguments against any eventual development of an ecological theory of time might include the observation that time is universally available to all organisms in any assemblage of co-occurring species, and cannot, therefore, be in limited supply. And if it is not in limited supply, then there cannot be competition for it, which weakens considerably its potential to act as a selective agent. Such counterarguments will be addressed, either directly or indirectly, in subsequent chapters as appropriate. Chapter 1, for instance, briefly reviews philosophical and cosmological theories of time, addressing questions of its existence, passage, and directionality. In doing so, the intent is both to challenge ecologists' preconceptions about time and its flow, and to thereby establish a foundation for thinking about time as more than simply a unidirectional arrow along which events and interactions unfold.

I have also tried to be comprehensive in my thinking about how to present parallels between time and other recognized biological resources, in hopes that this will bolster the argument for the consideration of time as a resource. These considerations will be presented in more detail in subsequent chapters, but here are the highlights. Like space, time may be available for use at many scales. The various scales at which time is available are recognizable as the units by which we measure it. Hence, during the progression of a particular reproductive season experienced by a long-lived organism, time may be available for use at scales of seconds, minutes, hours, days, and weeks, but not as years if the unit of a year exceeds the temporal scale of a reproductive season, even though time may have been allocated over the course of years to growth and development prior to reproduction.

Furthermore, owing to the paradoxically unidirectional and recurrent nature of time, the misuse of time at shorter scales of availability may, in long-lived organisms, be compensated for over scales that are unavailable during any single reproductive season, such as years. As well, some forms of time are intertwined with space and the presence or absence of other organisms. But there is one important difference between time and other resources, and it is this difference that might lend primacy to the role of time in ecology. Unlike other resources, the use of time may not render it unavailable for use by other organisms. However, its use by an organism for one purpose, its allocation to one life history stage or set of life history events, does in fact render that time unavailable to that same organism for allocation to other life history stages. Obviously, an organism may simultaneously allocate time to growth while flowering or gestating offspring, for instance. But the timing of the transition from one phenophase to the next in an organism's life cycle cannot be reversed or altered once that transition has been made. Hence, phenology represents not only a tracking of the availability of other resources through time; it also represents a strategy of allocating other resources to the use of time itself for growth, maintenance, and offspring production.

Before developing a treatment of phenology centered on a theoretical framework for the role of time in ecology, it might be a worthwhile exercise to reflect for a moment upon our own perception of time in a traditional ecological context. In other words, *what is time in ecology?* Generally speaking, time is considered as a conceptual axis, much like space, along which we can measure ecological events and their durations. Conveniently, cosmology defines events as occurrences in time and space (Hawking 1988, 1990). In ecology, time also allows us to describe, ascribe rates to, and quantify differences in, for example, changes in abundance within and among populations of single species and interacting species. It is used to quantify when events such as flowering times and other seasonal pulses of life history activity occur and to quantify changes in their occurrences in response to, for example, climatic warming. And so on. In such a framework, time is a measuring stick and half of the stage—the complementary half of which is space—upon which ecology plays out. As ecologists, we think we know what time is, and we know how to measure it as well as its ecological traces through the study of dynamics in many subdisciplines within ecology. But this knowledge is distinct from an understanding of the *role* of time in ecology. And no discipline is better suited to disclosing that role, and to the development and application of an ecological framework of time, than phenology.

Over the ensuing chapters, an argument will be constructed for the development of this theoretical framework. At the outset, however, I would like to present its main elements, while remaining cognizant of the fact that some elements of this list may be clear only in retrospect after more thorough treatment in the ensuing chapters. These elements are as follows: First, time is a biological resource in and of itself. Second, the evolutionary context of timing is best understood as it relates to *duration* of life history phases critical to survival and reproduction. Phenology is most commonly studied in the context of the timing of events, but to better understand variation in timing and what drives it, ecologists should place more emphasis on the influence of variation in timing on the duration of pheno-phases. The allocation of time, and other resources constrained by it, to critical phenophases or life history stages has adaptive value in the context of duration. Third, phenological stasis, advance, and delay can all be interpreted as strategies employed by the individual organism to optimize duration of, or the allocation of time to, crucial life history stages or phenophases related to growth, develop-ment, maintenance, and offspring production. Fourth, although individuals may compete for time, resource limitation and competition for time occur most clearly *within* the individual. Once allocated to a specific phenophase, time cannot be reallocated to another phenophase within the individual. Fifth, time used by the individual occurs as absolute cosmological time, recurrent time, and relative eco-logical time. Individual organisms use recurrent and relative ecological time to perpetuate their genes through cosmological time. In fact, the evolution of life history strategies that promote the use of recurrent and relative ecological time can be viewed as an elegant solution to a problem faced by all living organisms: contending with the irreversible, unidirectional, and inexorable passage of cos-mological time. Last, phenological patterns that emerge at the population, species, and community levels derive from the strategic allocation of time at the individual level. Before the argument for this framework is presented, however, let us begin, in the next chapter, with a brief examination of time itself.

CHAPTER ONE

What Is Time?

Presumably, ecologists are in agreement in assuming that time exists, that it flows, and that this flow has a definite and predictable direction. But perhaps we ecologists are also allied in wondering, at least on occasion, what time really is. It seems worthwhile, therefore, before proceeding under potentially false assumptions, that we address three questions related to the nature of time. First, does it in fact exist? Second, if so, does it flow or pass? And third, if it does flow, is it absolute and therefore gone when it passes, or does it recur? These questions will be essential in deciding whether time really can be considered a resource and a limited one at that.

PHILOSOPHICAL VIEWS OF TIME: IDEALISM, RELATIONISM, AND REALISM

The discipline of the philosophy of time offers insightful perspectives on the reality and nature of time, and this chapter will draw extensively on notable works in this field. Why should we review philosophical theories of time? Such treatments of time have been mostly anthropocentric, concerned with the reality and nature of time from human perspective. But if time is real, and represents a resource, then we must subsequently examine the nature of it from nonhuman perspective as well. Hence, examining what fields of study outside ecology have to say about the nature of time will aid us in developing a clearer understanding of the nature of time in ecology. As will be suggested in chapter 3, for instance, notions of the directionality of time may, in ecology, differ from the typical human experience.

Among philosophies of time, idealism denies the existence of time on the basis that change, an intuitive and apparently observable feature of time, is an illusion (Bardon 2013). A key feature of this perspective is the so-called paradox of movement: for an object to travel from one point to another, it must cross an infinite series of halfway points, which can never be achieved. Therefore, temporal idealism concludes that apparent movement, and change in general, is the misperception of an object occupying a space equal exactly to its own dimensions

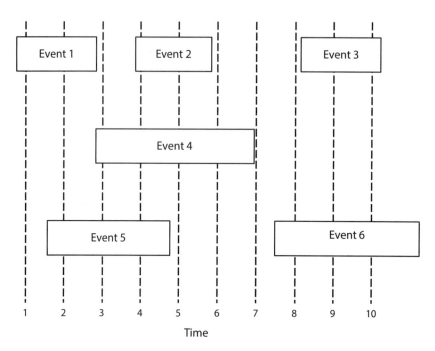

FIGURE 1.1. According to the relationist theory of time, time consists solely of events and their order, rank, and overlap with one another. Hence, in this example, time 2 consists of events 1 and 5, time 6 consists of event 4, and time 9 consists of events 3 and 6 (modified from Meyer 2013). This view of time is relevant to the concepts of relative ecological time, the phenological community, and phenological overlap among organisms in time that are discussed in subsequent chapters.

at any given instant. Furthermore, idealism contends that only the present is real. Any notion of past or future cannot be supported logically because neither is observable in the present, and if past and future do not exist, then the present cannot develop from the future or become the past, further refuting the reality of change. Relationism counters this by arguing that we should not conflate time and change because time is not a process but rather something independent of the process that merely allows us to measure it. In relationism, time consists fundamentally of events arranged according to their overlap, order, or rank with other events (figure 1.1) (Meyer 2013).

This view of time has relevance to the concept of the phenological community (chapter 3). It defines time according to subsets of events and their overlap or lack thereof, a notion that will be demonstrated in subsequent chapters as central to the interactions of individuals and species in time. Realism, last, simply represents the view that time is indeed real (Bardon 2013). Temporal realism has its most formal roots in Newtonian physics. Newton argued that the interdependence of time and space necessitates that time must be considered in absolute terms if we also consider motion in absolute terms (Newton 1687). This suggests that time exists independently of change, and that change occurs in time rather than as a result of the passage of time (Bardon 2013). This latter perspective is perhaps best represented by the cosmological or astrophysical view of time discussed later in this chapter.

A-SERIES, B-SERIES, AND THE FLOW OF TIME

If we accept that time does indeed exist, then we must next address whether time "flows" or "passes" and, if so, whether it does so unidirectionally and continuously (Callender 2011). This exercise is not simply esoteric or superfluous, but rather, for the purposes of developing an ecological framework for time, entirely necessary. The process of allocating time to biological maintenance, growth, and offspring production would, for instance, be very different if time were static or recurrent, and therefore unlimited, compared to such a process if time were in limited supply because of its unidirectional passage. Similarly, continuous passage of time might select for strategies relating to the allocation of time to maintenance, growth, and reproduction that could be expected to differ from strategies selected for under conditions of discontinuous passage of time. Ecology applies two different types of mathematical models in describing processes occurring in discrete time steps, such as population dynamics in species with nonoverlapping generations, and those occurring continuously, such as population dynamics in species with overlapping generations. However, ecology does not focus on the nature of time itself. Chapter 3 will suggest that there are different forms of time of relevance in ecological systems, but in all of these the assumption is that time itself is continuous. Nonetheless, we might regard one of these forms of time, relative ecological time (about which more will be said in chapter 3) as more continuous in largely aseasonal environments such as the tropics than it is in highly seasonal environments such as the Arctic.

Furthermore, if time does pass and if it does so unidirectionally, then in which direction does it flow? Our innate perception may be that the future lies ahead of us while the past lies behind us, but does this mean we move forward through time and thus that time washes backward over us? Causality, for instance, appears temporally asymmetrical: any action or decision in the present influences at least to some extent actions in the future but not those in the past, imbuing time with a sense of directionality (Callender 2011). Philosophers of time offer insights into such questions through the opposing theories of dynamic, or A-series, and static, or B-series, time. It may be tempting to dismiss philosophical theories of time

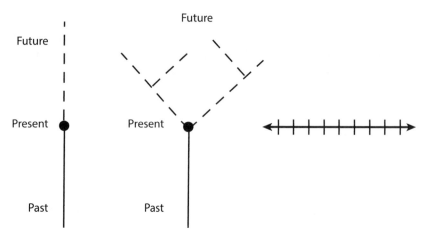

FIGURE 1.2. The A-series, or dynamic, theory of time, views time as progressing or flowing from past through present to some unknown future, which can be either determinate (left) or open (middle). The B-series, or static, theory of time (right) views time as consisting of events that are simply arrayed in an unchanging order that can be likened to a "wall calendar" view of time. Adapted from Hoefer (2011).

as irrelevant to the role of time in ecology because the former appear concerned mainly with human perception of time while the latter should operate universally and independently of human awareness. But astrophysics gives consideration to such questions as well, and does so from a perspective that is decidedly nonanthropocentric. Hence, there is no inherent reason for ecology to avoid such questions.

The A-series, or dynamic, theory of time assigns nonstationary temporal values to events. According to this theory, events move from being future events to present events, and then to past events (Hoefer 2011). Moreover, the designation of events as future or past can itself be variable and fluid in the A-series model. For instance, events can be far in the future or far in the past. And the same event can change from being far in the future to being in the near future, and then, after it has occurred, similarly move from being in the near past and eventually in the distant past (Bardon 2013). Within the A-series or dynamic theory of time, the present may be seen as moving toward a determined future (figure 1.2a) or an open future with many possible realizations or consequences of the present (figure 1.2b; Hoefer 2011). In astrophysical theories of time, summarized briefly in the next section, this latter view is consistent with the notion that the universe is asymmetrical with respect to time and information (Hawking 1996c; Penrose 1996a). Chapter 5 will suggest that in ecology, we can adopt such notions in our thinking of the manner in which the timing of events in, for example, a phenological sequence, influences the timing of subsequent events.

In contrast, the B-series, or static, theory of time views events as having fixed relations to one another, without notions of past, present, or future (Hoefer 2011). In two dimensions, we can visualize this as a number or event line (figure 1.2c), while in three dimensions this comprises the so-called block universe (Hoefer 2011; Bardon 2013). According to the static theory of time, the relations of events to one another in time do not change: an event is either before, simultaneous with, or after another event. Hence, the B-series theory of time conceptualizes events as existing in an unchanging order. According to this theory, events can be viewed in association with one another in time in much the same way that locations are viewed with respect to one another in space: hence, notions of *now* or *then*, and of *here* or *there*, are equally subjective (Bardon 2013). However, whereas relations among objects in space necessitate coexistence in time, relations among events in time do not necessitate coexistence in space. For instance, to observe that one object is beside, in front of, or behind another requires that they exist at the same time; in contrast, to observe that one event preceded, occurred simultaneously with, or followed another event does not necessitate that they did so in the same location (Meyer 2013). The static theory of time derives from the view that per-ceived change in the temporal relation of events is actually just an illusion and that there is no objective past, present, or future (McTaggart 1908; Price 2011).

SPACE-TIME AND ASTROPHYSICAL THEORIES OF TIME

The theory of space-time argues that space and time are inextricably interwoven, and as a consequence one cannot discuss processes unfolding, or dynamics occur-ring, in space without also considering such processes or dynamics in time, and vice versa (Minkowski 1909). Owing to the interdependent nature of space and time and their interaction in the continuum of space-time, the rate of passage of time, and perhaps even the order in which events unfold, depends upon relative spatial location or movement. Hence, the perceived rate of passage of time may depend upon the observer's state of motion or rest. This is perhaps best illustrated by the famous twin paradox, in which a twin sent into space on a near light-speed journey returns years later and is observed to have aged less than the identical twin who remained on the Earth. Moreover, whether an event is simultaneous with others, precedes others, or follows others depends upon the orientation of the space-time manifold (figure 1.3).

In figure 1.3, the order of events A, B, and C varies according to the observer's perspective in relation to the orientation of the axes of time and space: the events are either simultaneous or they occur in the order ABC or CBA without the events themselves having changed (Bardon 2013). This conclusion has relevance to

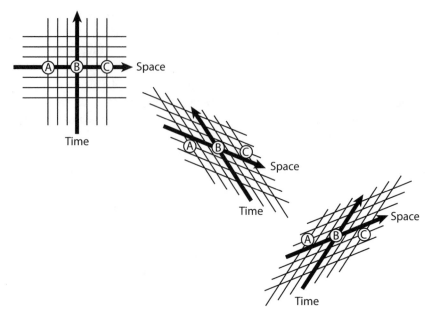

FIGURE 1.3. In the space-time manifold, the order of occurrence of events A, B, and C depends on the orientation of space-time. In the upper-left panel, all three events occur simultaneously. In the middle panel, event A precedes B, which precedes C. But in the lower-right panel, event C precedes B, which precedes A. Modified from Bardon (2013).

phenological dynamics in time and space that will become clear in the discussions of relative ecological time and the phenological community in chapters 3 and 5. For instance, whether the timing of expression of a given life history event is early or late from the perspective of the individual organism may depend not only on its absolute timing but also on its timing relative to that of the same trait by other individuals elsewhere in space or by the timing of expression of a trait by a member of another species with which that individual interacts.

Cosmological or astrophysical theories of space-time and the flow of time are inextricably linked to gravity and the second law of thermodynamics. Classical general relativity suggests that the perceived flow or directional passage of time may be a product of the action of gravity on the space-time manifold—in other words, the action of gravity on "the arena in which it acts" may have given time a beginning (Hawking 1996a). Sufficiently strong gravitational action may bend space-time and, thereby, alter the direction or flow of time. To illustrate this, Hawking (1996) used null geodesics to represent successive future-directed events in space-time, so-called event cones (figure 1.4). The points labeled p in figure 1.4 represent simultaneous events on a normal space-time surface resulting from

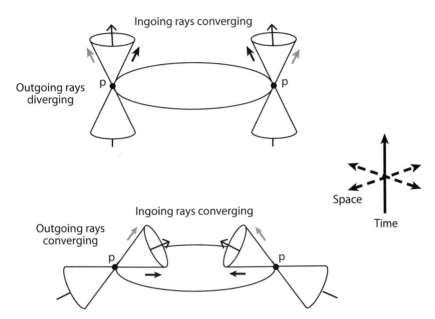

FIGURE 1.4. At a cauchy surface in space-time, outgoing rays of an event cone emanating from a present event will diverge (top), unless the cauchy surface represents a closed, trapped surface (bottom). In the latter case, sufficiently strong gravity can cause both outgoing and ingoing rays to converge and, potentially, slow or even stop the flow of time. Adapted from Hawking and Penrose (1996).

all possible events in the past (backward-directed light or event cones) and from which all possible future events emanate (forward-directed light or event cones).

In such a case, the outgoing rays of the paired event cones are divergent, while the ingoing rays are convergent (figure 1.4, top panel). However, on a closed trapped surface, such as during the collapse of a star, the gravitational field becomes sufficiently strong to pull the event cones inward (figure 1.4, bottom panel). In this case, both the outgoing and the ingoing rays become convergent, altering the perceived relation of past and future (Penrose 1996b). This suggests that, at least under conditions of sufficiently strong gravitational action, the directionality of time is not absolute (Hawking 1996b). Furthermore, in continuous Minkowski space-time, all points can be described as lying in the past of some future null infinity; however, near the singularity characteristic of a black hole, there can be said to be points that do not lie in the past of any future null infinity (Hawking 1996b). In chapters 3 and 5, the concepts of recurrent and relative ecological time will be developed, in which notions of phenological past and future become somewhat fluid.

The second law of thermodynamics dictates that entropy increases in the absence of an input of, or expenditure of, energy. Hence, the universe can be said to be asymmetrical with respect to entropy and order. Similarly, the universe can be described as asymmetrical with respect to time (Penrose 1996a), and as containing more information in its future than in its past (Hawking 1996b). This appears to relate to the possibility that the boundary conditions near the universe's past are smooth and regular, whereas those near its future are rough and irregular or chaotic (Hawking 1996c). It is this difference in the so-called Weyl tensor structure at the space-time topological boundaries of the universe that lends perceived unidirectionality to the cosmological arrow of time (Hawking 1996c). From the perspective of living systems, biological maintenance, growth, and offspring production represent the maintenance and creation of order, or reduction of entropy, which requires energy. As chapter 3 will argue, the allocation of time is the necessary element governing the use of, and transformation of, energy into biomass and offspring production.

Where, then, does the foregoing discussion leave us on the questions of whether time exists and, if so, what it really is? From a philosophical perspective, it is tempting to conclude that time is an experience, a concept applied to understand or describe change, rather than a thing in and of itself, just as speed or rate or direction all describe and allow us to understand motion, or just as distance allows us to describe and understand space (Kant 1781). This view may, at least in part, explain why ecology has struggled to recognize and formalize the notion of time as a resource because we, as ecologists, intuit that time is not necessarily a "thing" but rather a perception. However, cosmology and astrophysics appear to provide the most suitable interpretation of time for the purpose of understanding its role in ecology. A deceptively simple, yet profound, definition of time deriving from quantum theory suggests that time may be the amount of energy required to separate mass instances from a coincidence of two or more superposed alternative events (Penrose 1996a). This conceptualization is appealing because of its parallel to space as the separation of two or more potentially superposed positions. There are useful analogues in the treatment of space in ecology that can be applied in our thinking about time in ecology (chapter 3). And after the brief review of space-time theory in this chapter, it would seem grossly negligent to accept that space is a resource without simultaneously accepting that time is one as well.

CHAPTER TWO

Phenological Advance, Stasis, and Delay

Any discussion of phenology is implicitly linked to the subject of seasonality. When we read reports of unusually early arrival by migratory birds at breeding sites, of record early flowering by plants, or of delayed or prolonged onset of colorful leaf pigmentation in deciduous forests, we understand that these represent changes in the timing of biotic indicators of seasonality. The close association between phenology and seasonality is indicative of the biological basis of phenology. This association also contributes to the highly variable nature of phenological dynamics characteristic of some species but not others, and to variation in the direction and magnitude of changes in phenology within traits among species as well as within species among traits. For instance, as we will see later in this chapter, some taxonomic groups, such as amphibians, display strongly dynamic phenology (Todd et al. 2011), while others, such as some woody plants, display only weakly dynamic phenology (Root et al. 2003). Likewise, within some species of migratory birds the phenology of one trait in an annual life history cycle, such as egg-laying, may advance over several years in response to warming, but the phenology of another trait in its life cycle, such as migratory arrival, may remain fixed (Both and Visser 2001). But before proceeding with an examination of the complexity of phenology, we should first define *phenology* and distinguish it from *seasonality*.

PHENOLOGY VERSUS SEASONALITY

Organisms inhabiting temporally varying environments have evolved life history strategies aimed at optimizing the timing of resource acquisition and allocation to survival and reproduction (Roff 1992; Stearns 1992). The temporal scale at which these strategies are expressed by individuals of various species relates to the temporal scale over which resource availability varies. The temporal scale of variation in resource availability relates, in turn, to temporal variability in the availability of nutrients and minerals, and factors that constrain their availability, such as temperature or precipitation. The expression of a life history strategy related to resource acquisition for growth, survival, and reproduction under a given set of environmental conditions constitutes a life history trait. Variation through time in the expression of a given life history trait in relation to variation through time in environmental conditions constitutes phenology. This is not to be confused with life history *strategy*, although phenology is a component of life history strategies. For instance, some species, such as understory forbs in temperate deciduous forests, have a vernal life history strategy, in which they become active in the spring before canopy trees leaf out. This is a general description of the relative phenologies of two types of plants. However, the same forb species may advance their timing of activity in response to warming while trees in the same system do not. In this case, we would say that the phenology of the forbs has changed but that of the trees has not, even though their respective life history strategies have remained constant.

More properly, however, phenology is defined as the *study* of the occurrence of phenomena in relation to time. But the term is more commonly used in reference to the *phenomenon* of expression of life history events through time, as in the preceding paragraph. Throughout this book, it will be used in reference both to the discipline itself and to variation in the expression of life history events themselves. On occasion, it will also be used here in the somewhat redundant forms *phenological events* or *phenological dynamics*. In such instances, what are referred to are the discrete occurrence of life history–related phenomena in the former case, and fluctuations or trends through time in series of life history events in the latter.

Seasonality, in contrast, refers to temporal variation in abiotic environmental conditions. Most important among these, in the context of the evolution of life history strategies and the variation in their expression that constitutes phenology, are light, temperature, and precipitation or moisture. Seasonality on the Earth is a consequence of the obliquity of its angle to the Sun. Because of this obliquity, the photic regime, including day length and total incoming solar radiation (solar insolation), vary monthly from the equator to the poles, with the greatest variation occurring at high latitudes and the least at low latitudes (figure 2.1a,b). Moreover, within the tropics, solar irradiance varies seasonally with cloud cover and atmospheric moisture content. As consequences of the Earth's movement around the Sun and of its angle of obliquity to the Sun, average monthly temperatures increase from the poles to the equator, but their variability throughout the year diminishes from the poles to the equator (figure 2.1c). Hence, mean annual temperatures increase from the poles to the equator, and seasonality—the difference between mean annual minimum and maximum temperature—shows the opposite pattern, increasing from the equator to the poles (figure 2.1d).

Consequently, there is a distinct latitudinal gradient in photic or photoperiodic regime and temperature seasonality across the Earth. Equatorial and subequatorial regions represent relatively stable abiotic environments and high-latitude regions represent relatively dynamic abiotic environments on an annual basis. Importantly, however, despite relatively stable monthly temperatures, tropical regions experience seasonality on two temporal scales: intra-annual variation in precipitation and solar irradiance, and interannual fluctuation in drought and precipitation extremes, the latter driven mainly by the El Niño Southern Oscillation (ENSO). Nonetheless, considering the latitudinal variation in photoperiodic and thermal seasonality across the Earth, we should expect variability in phenological activity on an intra-annual basis to similarly increase with latitude. Hence, mid- and high-latitude systems are characterized by distinct growing seasons, while in tropical systems biological production occurs throughout the year. As well, because the magnitude of recent warming has been greater at high latitudes than at low latitudes (IPCC 2014; Post et al. 2018), phenological advances should increase in rate or magnitude from low to high latitudes.

We should expect such patterns on the basis of an understanding of the role of environmental seasonality in the evolution of life history strategies related to survival and reproduction. Organisms in highly seasonal environments, such as those at high latitudes, experience a seasonal cycle of resource abundance and dearth. As latitude increases, so does the relative difference in resource availability between seasonal abundance and seasonal dearth. As well, as latitude increases, the seasonal window of resource abundance shrinks, while the rates at which resource availability increases and declines seasonally increase. Hence, organisms in such highly seasonal environments can be expected to have evolved life history strategies that facilitate seasonal resource acquisition and that themselves are expressed in a highly seasonal manner. Such strategies may consequently be highly responsive to seasonal changes in temperature or day length, two important abiotic constraints on the seasonal expression of life history traits in high-latitude environments. Those traits governed by and responsive to seasonal changes in day length are endogenous, and their expression is unlikely to vary interannually. In contrast, those traits governed by and responsive to seasonal changes in temperature should have the capacity to vary in their expression interannually. This is one reason we should expect trends in phenological dynamics associated with recent climate change to be more pronounced at higher latitudes. But what do the data on long-term variation in phenology tell us?

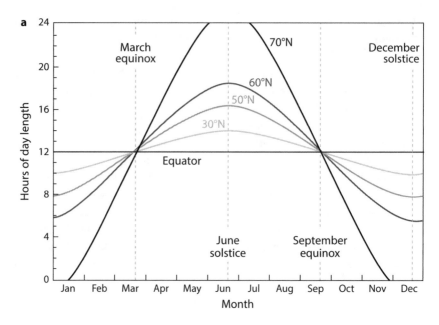

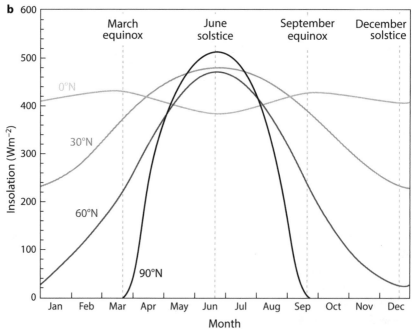

FIGURE 2.1. Variation in seasonality with latitude across Earth, exemplified by greater seasonal variation in day length, incoming solar radiation, and temperatures at high versus low latitudes. (a) Monthly variation in the number of hours of daylight in a 24-hour period at the equator and

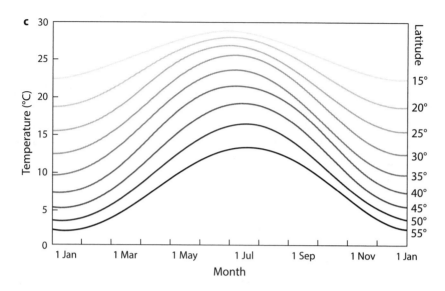

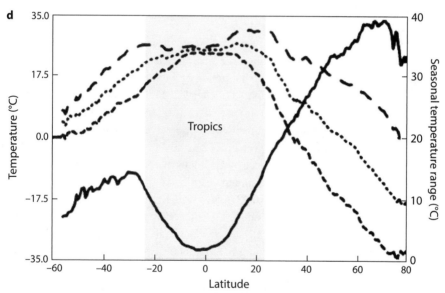

latitudes 30, 50, 60, and 70 degrees N. Opposing patterns apply in the southern hemisphere. (b) Similarly, monthly variation in incoming solar radiation at latitudes 0, 30, 60, and 90 degrees N. As in panel (a), opposing patterns apply in the southern hemisphere. (c) Variation in monthly mean temperatures in the northern hemisphere with latitude. (d) Patterns of variation in annual minimum, mean, and maximum temperatures with latitude (dashed and dotted lines, left y-axis), and in seasonal temperature range with latitude (solid line, right y-axis). Panels a and b are reproduced from chapter 6, "Energy and Matter," physicalgeography.net. Panel (c) is reproduced from Wilczek et al. (2010), and panel (d) is reproduced from Wright et al. (2009).

UBIQUITOUS VARIABILITY IN PHENOLOGICAL TRENDS

This section presents an overview of the central findings of four major reviews of phenological change over the past several decades. The three earliest of these reviews (Walther et al. 2002; Parmesan and Yohe 2003; Root et al. 2003) all appeared in the journal *Nature* within ten months, and two of these appeared in the same issue (Parmesan and Yohe 2003; Root et al. 2003). In total, and as of this writing, these three reviews have been cited, according to Google Scholar, more than 19,000 times. All three of these early reviews figured prominently in the Fourth and Fifth Assessment Reports of the Intergovernmental Panel on Climate Change (Pachauri and Reisinger 2007; IPCC 2014), and were included in a separate major review of springtime phenological responses to climate change (Badeck et al. 2004). Two of them (Parmesan and Yohe 2003; Root et al. 2003) were the focus of a subsequent review of factors explaining variability in estimates of phenological changes in response to climatic warming (Parmesan 2007). The fourth major review included in this brief overview derives from the report by Working Group II in the IPCC's Fifth Assessment Report (IPCC 2014), and includes more recent individual studies.

The first of these three early reviews reported advances in the annual timing of onset of flowering and leaf unfolding by numerous species of plants in Europe and North America, advances in the annual timing of spring appearance by 18 species of butterflies in the United Kingdom, and advances in the annual timing of spring migration and breeding by numerous species of birds in Europe and North America (Walther et al. 2002). Midrange estimates of rates of advance in the annual timing of flowering and leaf out varied between approximately −1.5 to −2.3 days per decade, while the timing of spring appearance by UK butterflies advanced by approximately −3 days per decade, and the annual timing of spring migration and breeding by birds advanced by approximately −2.9 to −3.4 days per decade (figure 2.2).

Root et al. (2003), using a taxonomically and geographically more extensive set of metadata, presented estimates of rates of advance in the annual timing of spring emergence of invertebrates, breeding in amphibians, spring migration and laying dates in birds, and flowering and leaf unfolding in plants. Across all taxa, the mean rate of advance in spring phenology was estimated at approximately −5 days per decade, with the greatest rate of phenological advance occurring in birds and the lowest rate in trees (figure 2.3; Root et al. 2003). Root et al. (2003) also partitioned their estimates into two bands of degrees north latitude to test the hypothesis that phenological advances at higher latitudes have exceeded those at lower latitudes. In support of this hypothesis, advances in the latitudinal range 50 degrees N to 72 degrees N (−5.5 ± 0.1 days per decade) were significantly greater than those in the latitudinal range 32 degrees N to 49.9 degrees N (−4.2 ± 0.2 days per decade).

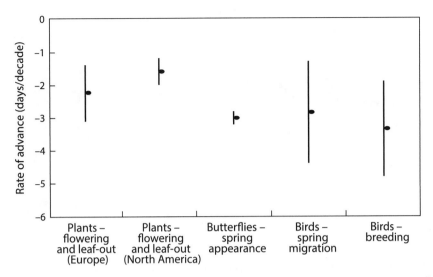

FIGURE 2.2. Variation in rates of phenological advance (days per decade) among plants, butterflies, and birds based on a review of published estimates compiled by Walther et al. (2002). Points represent midrange estimates, while bars below and above represent minimum and maximum estimates, respectively. While butterflies exhibit the lowest level of variation in phenological rates of advance in this example, birds display the greatest. Adapted from Walther et al. (2002).

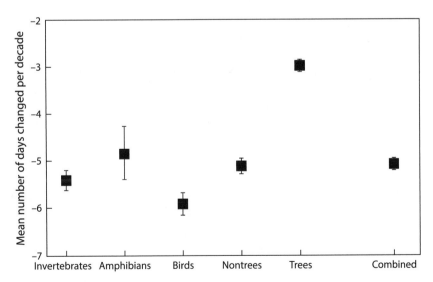

FIGURE 2.3. Mean (±1 standard error) rates of phenological advance (days per decade) for events pooled within taxa for invertebrates, amphibians, birds, nontree plants, trees, and for all taxa combined. Adapted from Root et al. (2003).

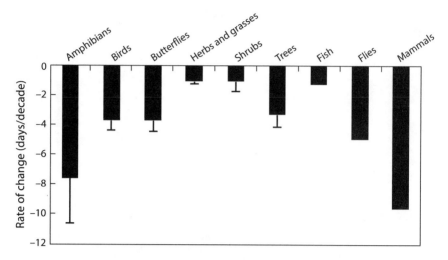

FIGURE 2.4. Mean (±1 standard error) rates of phenological advance (days per decade) for events pooled within taxa for amphibians, birds, butterflies, herbs and grasses, shrubs, trees, fish, flies, and mammals. Modified from Parmesan and Yohe (2003) and Parmesan (2007). Note that in the meta-analyses highlighted in figures 2.2 and 2.3 and this figure, reported phenological trends are all negative. Emphasis on examining positive phenological trends—that is, delays in onset of phenological events—finally appears in Parmesan (2007).

Using largely the same source data as Root et al. (2003), Parmesan and Yohe (2003) estimated the grand mean rate of advance in springtime phenological events across all taxa at approximately −2.3 days per decade, or nearly half the rate of advance reported in Root et al. (2003). The difference in the two estimates has been ascribed to disparate data selection criteria between the two studies (Parmesan 2007). A subsequent reanalysis of a combined set of data from both Root et al. (2003) and Parmesan and Yohe (2003) presented taxon-specific estimates of rates of advance in the annual timing of spring phenological events as well as the global mean rate of advance across all taxa (Parmesan 2007). This reanalysis revealed the greatest rate of phenological advance had occurred in amphibians, while lowest rates of advance had occurred in herbs and grasses (figure 2.4); the grand mean rate of phenological advance estimated across all taxonomic groups was −2.8 days per decade (Parmesan 2007). This meta-analysis also detected a significant increase in the rate of phenological advance of approximately −0.18 days per decade with each one-degree increase in degrees north latitude (Parmesan 2007).

Both Root et al. (2003) and Parmesan and Yohe (2003), and the subsequent synthesis meta-analysis by Parmesan (2007) of data sets used in both of those studies, also examined lack of phenological trends and trends that were inconsistent with

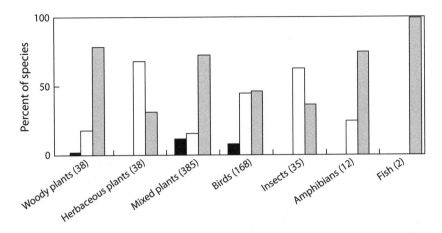

FIGURE 2.5. Percentage of the total number of species in several taxonomic groups examined by Parmesan and Yohe (2003) that exhibited significantly positive phenological trends (that is, phenological delays; black bars), no significant phenological trends (that is, phenological stasis; white bars), or significantly negative phenological trends (that is, phenological advances; gray bars). The total numbers of species examined within each taxonomic group are reported in parentheses. Adapted from Parmesan and Yohe (2003).

expected shifts in phenology in response to climatic warming—that is, phenological stasis and delays. Root et al. (2003) found evidence of nonstatistically significant advances in phenology that ranged from −1 day per decade to −24 days per decade, and nonsignificant delays in phenology that ranged from 2 to 6 days per decade. Overwhelmingly, however, shifts in phenology were negative, indicating a general pattern of advancing phenology across taxa in concert with the past century of warming (Root et al. 2003). Parmesan and Yohe (2003) provided taxonomic resolution in their assessment of the proportion of the total number of phenological shifts that were consistent with versus opposed to expected shifts in response to climatic warming. Their analysis revealed that most taxonomic groups examined showed no phenological shifts inconsistent with expected responses to climatic warming. But, of those that did, between 2.6 percent and 12 percent of species exhibited delayed phenological dynamics—that is, positive trends in timing. The highest percentage of these occurred among mixed plant species (figure 2.5; Parmesan and Yohe 2003). In contrast, all taxonomic groups except fish included species for which no significant change in phenology was detected. These percentages were considerably higher than those exhibiting delayed phenological dynamics, ranging from 18 percent to 68 percent of species, with the highest rates of phenological stasis occurring in herbaceous plants and insects (figure 2.5; Parmesan and Yohe 2003).

The review of recent phenological changes compiled by Working Group II in the Fifth Assessment Report of the IPCC (WGIIAR5) relied considerably upon results published in Root et al. (2003), Parmesan and Yohe (2003), and Parmesan (2007), but added information from publications appearing since those three studies (IPCC 2014). Included among these were results from recent studies providing estimates of regional scale changes in the timing of onset of the plant growing season deriving from satellite observations. The great value of such studies is the perspective they offer of dynamics unfolding over tremendous spatial extents. However, comparisons of the estimates they provide to those obtained from ground-based observations are problematic because satellite observations are largely blind to species-level dynamics (Schwartz 1998). Comparisons among satellite-derived estimates themselves can also be problematic owing to variation in instrumentation and methodology through time (IPCC 2014). Nonetheless, the WGIIAR5 review reported an average rate of advance in the timing of the annual onset of plant growth on land across the northern hemisphere (latitude 30 degrees N to 80 degrees N) of −2.9 days per decade between 1982 and 1999, and a rate of advance of −0.2 days per decade since then (IPCC 2014). As well, the annual timing of the end of the plant growing season on land over the northern hemisphere was delayed at a rate of 2.4 days per decade between 1982 and 2008. Together, these latter two observations suggest that the length of the annual plant growing season has increased across the northern hemisphere in association with warming over the past three decades (Pachauri and Reisinger 2007; IPCC 2014). Owing to the lack of species-level resolution in such satellite data (Schwartz 1998), however, it is impossible to conclude on this basis alone whether individual plant species have undergone both advanced early-season phenology and delayed late-season phenology.

Working Group II of the IPCC's Fifth Assessment Report also touched briefly upon recent phenological shifts in animal species. Relying once again on the review by Parmesan (2007), it noted that the annual timing of egg-laying has advanced across 41 species of birds at a mean rate of approximately −3.7 days per decade (IPCC 2014). The annual timing of parturition by red squirrels in a study conducted at Kluane Lake, Canada, was also reported to have advanced at a remarkable rate of −18 days per decade (Reale et al. 2003). Subsequently, discussion of recently documented phenological trends in animal species in the AR5 is limited to descriptive accounts absent quantitative presentation of rates of change. These include advanced timing of emergence from hibernacula by yellow-bellied marmots in the Rocky Mountains of Colorado (Ozgul et al. 2010); delayed emergence from hibernacula by Columbian ground squirrels in Alberta, Canada (Lane et al. 2012); and the observation that timing of spring arrival by short-distance migratory bird species has advanced to a greater extent than has that of long-distance migrants (Saino et al. 2009). Additionally, WGII of the AR5 noted that several recent studies have documented shifts in phenology in amphibians and insects, though these were not detailed in the report. In the following section, quantitative estimates of rates of phenological advance or delay described in the AR5, and drawn from the primary literature, will be presented for the interesting and informative patterns they reveal when combined with the major reviews addressed earlier.

But, for now, what does this overview tell us? First, it tells us that, in general, recent phenological trends in plants and animals tend overwhelmingly to be negative—that is, species across a diverse array of taxa appear to have tended toward earlier timing of springtime activity in concert with recent climatic warming. Second, however, and perhaps more interestingly, it tells us that there has not been a universal tendency toward earlier timing of springtime phenological events with warming. A substantially nontrivial proportion of case studies examined in these reviews has revealed nonsignificant changes in the timing of events, and, in some instances, delayed timing of events. Evidently, therefore, some species are either incapable of advancing their phenologies while those with which they co-exist in the same species assemblages are advancing theirs, or they employ a distinct strategy of phenological stasis or delay while others advance (*sensu* Todd et al. 2011). Similarly, a comprehensive analysis of trends in springtime phenological events spanning 12 taxa, including six species of plants, three species of birds, two species of insects, and one species of amphibian recorded at 12 to 160 sites across Japan and South Korea between 1953 and 2005, reported both advances and delays in phenology (Primack et al. 2009). Across the plant species examined, flowering and leaf-out phenology trended toward earlier onset, with species at two-thirds to three-fourths of all sites showing negative trends (Primack et al. 2009). In contrast, two of the bird species, both insect species, and the lone amphibian examined all trended toward later occurrence at a majority of sites (Primack et al. 2009). And, as noted in the introduction, additional examples of phenological delay directly or indirectly associated with warming have been documented in an array of taxa. Much of the rest of this book will be devoted to developing an understanding of the reasons for such divergent patterns in phenology. This understanding should exceed simply ascribing such differences to variation among species in the particular abiotic constraints on the expression of their respective life history strategies.

Last, this overview tells us that there is, indeed, some evidence, albeit slight up to this point, for a latitudinal trend in rates of phenological advance. This evidence appears limited primarily to three lines of support, despite its common assertion in the literature: the binned analysis by Root et al. (2003), which revealed a rate of phenological advance of approximately −1.3 days per decade greater at high compared to midlatitudes; the regression analysis by Parmesan (2007) across 203 trends, which revealed an increase in the rate of phenological advance of approximately −0.18 days per decade with each degree increase in north latitude; and the observation that some of the greatest rates of phenological advance yet reported in the literature derive from the site monitored at the highest latitude to date, where the mean rate of advance in springtime events across 21 taxa was approximately −14.5 days per decade (Høye et al. 2007). In the next section, we will take a much closer look at patterns in phenological trends comprising the studies reviewed in this section, including variation across taxa and latitude.

PHENOLOGICAL DYNAMICS ACROSS TAXA, LATITUDE, AND TIME

To examine general patterns in phenological trends across taxa and life history traits, consider an analysis of metadata from 76 original sources cited in the major reviews discussed earlier, as well as more recent sources that published phenological trend estimates based on at least ten years of data (Post et al. 2018). These included 743 unique estimates of phenological trends, accompanied by taxonomic identifiers, and, in most cases, some indication of statistical significance, start and end years of the time series, number of years the time series comprised, and latitude.

Overall, approximately 60 percent (449) of the trends in phenological dynamics that reported significance tests in this metadata pool were significant at the $\alpha = 0.05$ level. Hence, nearly 40 percent (294) of reported trend estimates were non-significant. This nonstatistical significance might in some cases be attributed to sample size, or time series length (Post et al. 2018). In a strict sense, however, it indicates that such trend estimates do not differ from zero, suggesting that phenological stasis is not uncommon among the studies represented in this meta-analysis. Birds dominated the numbers of both significant and nonsignificant trends reported, and were followed by plants, invertebrates, amphibians, mammals, and fish (figure 2.6). For birds and amphibians, the numbers of nonsignificant trends reported were nearly equal to the numbers of significant trends reported (figure 2.6). No nonsignificant trends were reported for mammals or fish, and the numbers of significant trends reported for these two groups were five and two, respectively.

Studies upon which phenological trend estimates were based commenced as early as 1928 and as late as 2002, and on average began in 1967. These values were identical for the pool of significant and nonsignificant trends and for significant trends alone (figure 2.7a). Although the mean start date differed by only

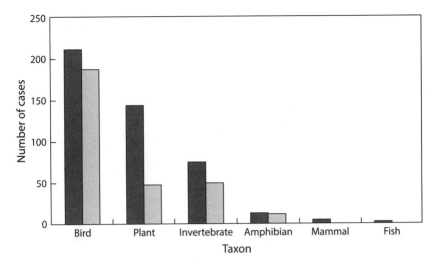

FIGURE 2.6. Numbers of cases included in the meta-analysis of taxon-specific and trait-specific phenological trends (Post et al. 2018) reviewed in this chapter. Black bars represent the numbers of significant ($P \leq 0.05$) phenological trends analyzed for each taxonomic group. Gray bars represent the numbers of nonsignificant ($P > 0.05$) trends. Determination of significance was based on P-values reported in the source literature.

3 days between studies demonstrating significant versus nonsignificant trends in phenology, the earliest study demonstrating nonsignificant trends began in 1936, or 8 years later than the earliest study documenting significant phenological trends (figure 2.7a).

The numbers of data points upon which phenological trend estimates were based (that is, the lengths of the time series), varied from 5 to 61 years across all reported trends (figure 2.7b; Post et al. 2018). This is not the same as the span of years covered by the data, for in many instances the time series were discontinuous. The minimum, mean, and maximum time series length did not vary substantially between significant and nonsignificant trends (figure 2.7b).

The full range of estimates of rates of phenological change across both significant and nonsignificant trends varied from an advance of −51.3 days per decade to a delay of 26.3 days per decade (figure 2.7c). This range was, however, determined by the maximum and minimum estimates of nonsignificant trends, which were identical to those for the pooled set of significant and nonsignificant trends. By contrast, statistically significant trends varied between a maximum rate of advance of −36.3 days per decade and a maximum rate of delay of 7 days per decade, with a mean rate of −5.68 days of advance per decade (figure 2.7c; Post et al. 2018).

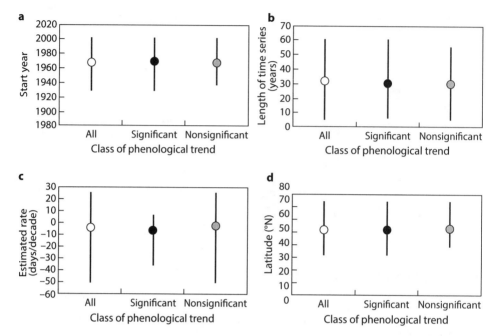

FIGURE 2.7. Composition of the pool of phenological trends used in the meta-analysis (Post et al. 2018) reviewed in this chapter according to (a) the start year of the original studies upon which estimates of phenological trends were based; (b) the length of the phenology time series (in years) upon which the original analyses were conducted; (c) the originally published estimates of phenological trends; and (d) the latitude in degrees north of the phenology time series used in the original studies. In each panel, the dots represent the means for the pooled trends (white), significant trends (black), and nonsignificant trends (gray), and the bars represent the ranges (minimum and maximum values).

The geographic range covered by the studies upon which phenological trend estimates were based varied from 31.9 degrees N latitude to 74.5 degrees N latitude (figure 2.7d). The mean latitude of study was significantly farther south for significant (51.3 degrees ± 0.4 degrees N latitude) versus nonsignificant (53.4 degrees ± 0.5 degrees N latitude) trends (Post et al. 2018; see figure 2.7d). This difference was likely driven by the fact that studies demonstrating a significant trend in phenology were conducted as far south as 31.9 degrees N latitude, while studies demonstrating no significant trend in phenology were conducted at a minimum latitude of 38.9 degrees N, and both types of studies extended as far north at 74.5 degrees N latitude (figure 2.7d).

When we look more closely at the actual estimates of phenological trends, several interesting patterns emerge. Most notable among these are obvious and strong patterns of heteroscedasticity. When both nonsignificant and significant (at the

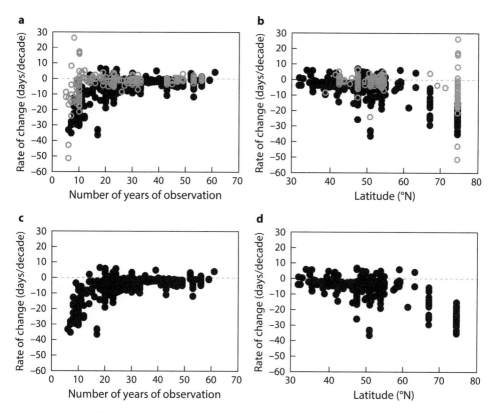

FIGURE 2.8. Results of the meta-analysis of published phenological trends (Post et al. 2018) reviewed in this chapter. Panels (a) and (c) depict the relationship between estimated rate of phenological change (days per decade) and length of the time series (in years) of the original study. Panel (a) shows both significant (black dots) and nonsignificant (gray circles) estimates of phenological trends, while panel (c) shows only significant trends. Panels (b) and (d) depict the relationship between estimated rate of phenological change and latitude (degrees north) of the original study. Panel (b) shows both significant and nonsignificant estimates of phenological trends, while panel (d) shows only significant trends. In all panels, negative trends represent phenological advances, and positive trends represent phenological delays. Modified from Post et al. (2018).

0.05 level) trend estimates are plotted against the length of the time series upon which they are based, there is a pronounced decrease in the variability among estimates in the rate of phenological advance with an increase in the number of years upon which the estimates are based (figure 2.8a; Post et al. 2018). For this metadata pool, the number of years upon which estimates of rates of phenological advance were based varied between 5 and 61. Among the shortest time series, estimates of rates of change varied from an advance of approximately −50 days per decade to a delay of approximately 30 days per decade, neither of which was significant (figure 2.8a). At the opposite end of the spectrum, estimates of rates of change deriving from the longest time series varied between an advance of approximately −1 day per decade to a delay of approximately 5 days per decade (figure 2.8a). Notable heteroscedasticity is also evident in a plot of trend estimates versus the latitude at which studies upon which they are based were conducted (figure 2.8b). This pattern of increasingly variable trend estimates the farther north one looks appears driven by the large range of nonsignificant estimates deriving from the farthest northern study in the metadata pool (figure 2.8b), which, coincidentally, is also one of the shortest studies in that pool.

Removing nonsignificant trends from the metadata pool, two additional patterns emerge. First, there is an obvious negative skew in the trend estimates toward strong rates of advance among the shortest time series (figure 2.8c; Post et al. 2018). In fact, positive trends, indicating phenological delays, are absent from the metadata pool among time series shorter than approximately 15–20 years (figure 2.8c). Second, we see a negative association between rate of phenological advance and latitude that is consistent with the conclusions of the bin analysis in Root et al. (2003), the meta-analysis in Parmesan (2007), and the conclusions of WGIIAR5 (IPCC 2014) (figure 2.8d; Post et al. 2018).

The pattern of heteroscedasticity that is evident when significant and nonsignificant trend estimates are plotted against time series length (figure 2.8a) suggests, quite unsurprisingly, that longer studies may provide greater precision in estimates of rates of phenological change. But why do shorter time series tend to produce statistically significant estimates of the most pronounced rates of phenological advance? One explanation lies in sample size: short time series, when they do produce significant correlations, tend to have steep slopes. This does not, however, explain why those slopes are, in this metadata pool, universally negative. Another pattern in the metadata may help explain this association.

There is a clear, negative association between latitude and length of phenological time series in this metadata pool (figure 2.9a; Post et al. 2018). This pattern indicates that the shortest studies have, on average, been conducted at the farthest northern sites. In part, this may be explained by the history of citizen science and phenological record keeping. It might also reflect a pattern of geographic diffusion to higher latitudes of phenological investigations related to climate change research. Further support for this explanation may be lent by the positive association between the first year of observation in any particular phenological record in this metadata pool and the latitude at which that record has been compiled (figure 2.9b; Post et al. 2018). This pattern indicates that, in general, far northern studies of phenology have their origin in the 1980s, 1990s, and early 2000s. This is also the period when public and scientific awareness of global climatic warming began and through which it has grown. Finally, there is

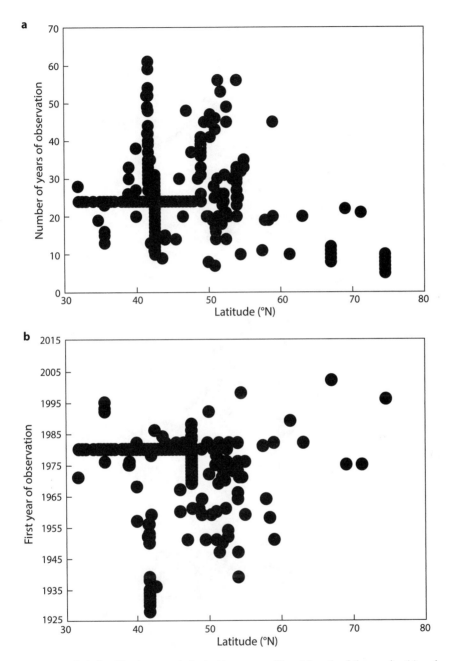

FIGURE 2.9. Relationships between latitude (degrees north) and length of time series (a) and start year of time series (b) of the original studies from which estimates of phenological trends used in the meta-analysis (Post et al. 2018) reviewed in this chapter were derived. High-latitude studies of phenological dynamics, where rates of phenological advance have been greater (see figure 2.8d), have been shorter and more recent than those conducted at lower latitudes. Modified from Post et al. (2018).

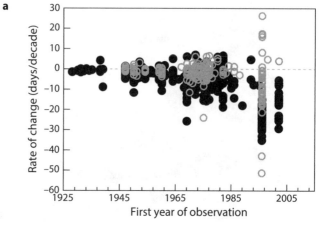

a

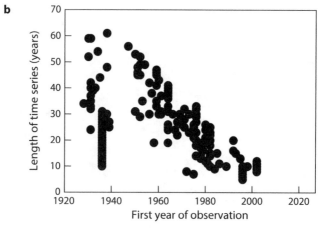

b

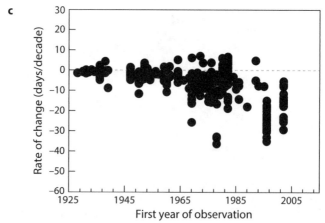

c

a third pattern of heteroscedasticity in the pooled significant and nonsignificant trend estimates. This pattern suggests that observational investigations of phenological dynamics initiated more recently have tended to produce more highly variable estimates of phenological trends (figure 2.10a). Such an association may be assumed, and logically so, to be due at least in part to the fact that more recently initiated studies of phenology have, by necessity, also been among the shortest (figure 2.10b). However, there is also a fairly strong, negative association between statistically significant rates of phenological advance and the first year of observation upon which the time series from which they derive are based (figure 2.10c). In other words, studies begun more recently have reported stronger trends of phenological advance than have studies begun further in the past. In fact, in this metadata pool, no statistically significant phenological delays have been detected in studies initiated since the mid-1990s. This last pattern evidently conflates latitude (these studies are also the farthest north) and recent warming, which has been most pronounced over the past two to three decades, as we will see later (Post et al. 2018).

Last, there is the matter of variation among taxonomic groups, and the observed dynamics in life history traits specific to them, in estimates of phenological trends. Both Root et al. (2003) and Parmesan (2007) noted differences in mean rates of phenological advance among plants and animals, and among different groups within plants and animals. Such variation is also evident in the pooled metadata from those studies and the more recent studies cited in the IPCC AR5. Examining trend estimates pooled across life history traits but separated by taxa, we note that significant trends in the timing of spring events were strongest in amphibians and invertebrates, which have undergone mean advances of -13.2 ± 1.9 and -10.3 ± 0.8 days per decade, respectively (figure 2.11a). In contrast, trends are nearly equivalent among mammals, birds, and plants in general (figure 2.11a). Among significant trends only, the greatest proportion of negative trends, indicative of phenological advance, occurred among birds, while the greatest proportion of positive trends, or phenological delays, occurred in plants (figure 2.11b).

FIGURE 2.10. Relationships between the first year of observation (the start year) of the original study from which estimates of phenological trends used in the meta-analysis reviewed in this chapter (Post et al. 2018) were derived, and (a) estimates of both significant (black dots) and nonsignificant (gray circles) phenological trends; (b) the length of the time series (in years) upon which estimates of phenological trends were based; and (c) estimates of significant phenological trends only. More recent studies of phenological dynamics have been shorter, and they have produced stronger estimates of rates of phenological advance, than earlier studies. Modified from Post et al. (2018).

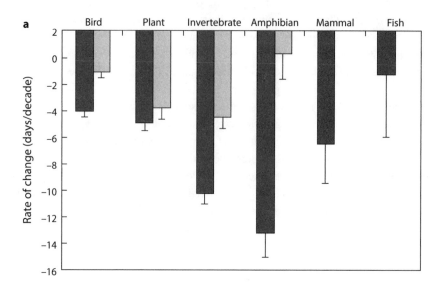

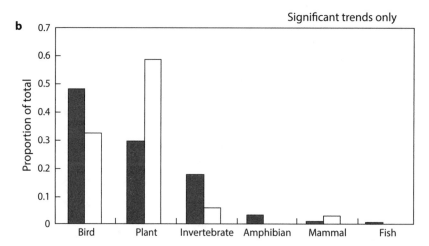

FIGURE 2.11. (a) Taxon-specific mean (±1 standard error) rates of phenological advance (days per decade) resulting from the meta-analysis reviewed in this chapter (Post et al. 2018). Black bars represent the means of significant phenological trends, and gray bars represent the means of nonsignificant trends. Determination of significance of trends was obtained from the original sources in which trend estimates were reported. (b) Proportions of the total number of significant phenological trends in panel (a) that are negative (solid bars) or positive (open bars). Derived from Post et al. (2018).

Further dissecting these metadata according to taxon-specific traits, the least resolution of phenological trends to the level of specific life history traits is apparent in amphibians, fish, and invertebrates (figure 2.12). In amphibians, only timing of breeding and timing of first spring appearance are represented in the metadata pool, and of these, timing of breeding has advanced more rapidly (figure 2.12a). In birds, trends in annual timing of arrival on breeding grounds, and in annual timing of breeding, egg-laying, hatching, and spring and fall migration are represented. Among these, the most rapid rates of statistically significant advance have occurred in arrival, egg-laying, and hatching, while nonsignificant trends are present for those three traits and timing of breeding (figure 2.12b). In fish, the only trait represented in the metadata pool is the annual timing of spawning, which has advanced significantly at a very modest rate (figure 2.12c). Among invertebrates, both the annual timing of spring emergence and of peak biomass have advanced significantly, while nonsignificant trends in the timing of spring emergence are also evident (figure 2.12d). Among mammals, the timing of parturition has advanced most rapidly, but significant trends were also present in the timing of spring emergence from hibernacula, timing of spring appearance of predators in avian nest boxes, and in the annual timing of weaning (figure 2.12e). Last, plants displayed the greatest significant mean rates of advance in the annual timing of spring events, with significant rates of advance in timing of budding, flowering, fruiting, leaf-out, mixed leaf-out/flowering, and pollination (figure 2.12f). The annual timing of fall senescence was significantly delayed on average, while flowering and leaf-out also displayed nonsignificant trends (figure 2.12f).

At this point, we may be left wondering what these patterns can tell us and why it is important to address them. My purpose with this examination of variation in phenological trends with taxon, space, and time is ultimately related to the question of how and why rates of phenological change associated with climatic warming vary with latitude. But before we get to that, it is safe to say, at this point, that there is ample evidence for recent advances in the timing of spring life history events across an array of taxonomic groups. It is just as safe to say that many time series do not indicate significant phenological advances in multiple traits and taxa, and some evidence for significant phenological delays. Now we are left to explain the interesting observation that the annual timing of expression of life history traits related to production, survival, and reproduction has not advanced in concert with recent warming equally among taxa, equally within taxa, nor equally among or within traits. These points will be returned to in the following chapter.

The beginning of this chapter highlighted the fact that abiotic environmental seasonality increases from the equator to the poles. This suggests that temperature limitation of the timing of expression of life history traits related to growth, survival, and reproduction should be stronger at high than at low latitudes. From this,

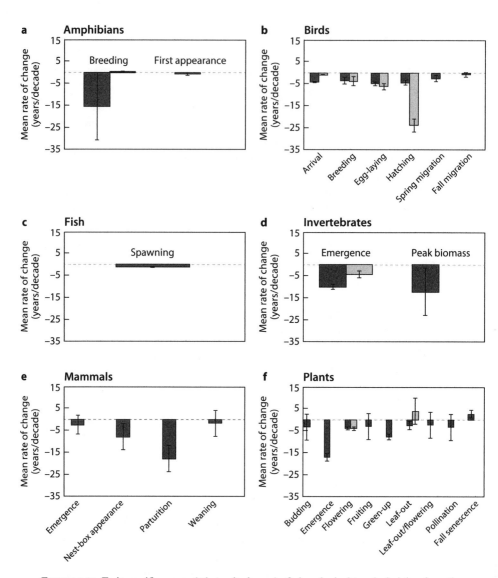

FIGURE 2.12. Trait-specific means (±1 standard error) of phenological trends deriving from the meta-analysis reviewed in this chapter (Post et al. 2018) for (a) amphibians, (b) birds, (c) fish, (d) invertebrates, (e) mammals, and (f) plants. Black bars represent means of significant phenological trends, and gray bars represent means of nonsignificant trends. Determination of significance of trends was obtained from the original sources in which trend estimates were reported.

it also follows that alleviation of temperature constraints on the expression of such life history traits should be greater at high than at low latitudes because the rate of warming has been greater at high than at low latitudes. But after examination of the patterns highlighted in this section, it should be evident that quantification of change in rates of phenological advance with latitude is problematic for several reasons. Hence, quantifying this relationship requires controlling for several factors, including time series length, the year in which observations were begun, taxon, and the specific traits observed.

This can be achieved by applying a generalized additive modeling (GAM) approach (Hastie and Tibshirani 1999). This approach allows us to examine relationships between continuous numeric predictor and response variables, while simultaneously accounting for variation in the response variable owing to nominal categorical predictors, such as taxon and trait. Having undertaken this analysis, we find that the negative association between rate of phenological advance and latitude remains robust and statistically significant (Post et al. 2018). Moreover, we can see that the estimate of the magnitude of change in the rate of phenological advance with latitude, the slope of the relationship between these two variables indicates that the rate of advance increases by approximately 0.4 to 0.5 day per decade with each degree north latitude gained (Post et al. 2018). This estimate is approximately 2.5 times greater than that deriving from the meta-analysis in Parmesan (2007).

VARIATION IN RATES OF LAND-SURFACE WARMING ACROSS LATITUDE AND TIME

We now have a basis for comparison of this pattern to that of variation in the decadal rate of warming with latitude. Such a comparison should indicate whether the increase in rates of springtime phenological advance with latitude is consistent with any such association between the rate of warming and latitude. This, in turn, should indicate whether springtime events are advancing at greater rates at higher latitudes simply because the rate of warming has been greater there than at lower latitudes. Alternatively, it should indicate whether springtime events are occurring earlier or later than expected farther north based on the rate of warming. It is this deviation from expectation that would inform the biological basis for variation in rates of phenological change within taxa, among taxa, and across space and time.

As a means of addressing this issue, let us consider latitudinal variation in rates of warming. But rather than doing so in a temporally static framework, let us employ an approach that considers how such variation itself varies with the period of consideration. A brief aside is warranted here to justify this approach.

The meta-analysis of phenological trends across taxa revealed, among other patterns, that high-latitude phenological data sets tend to be more recent and of shorter duration than those deriving from studies at lower latitudes (see figure 2.9). Hence, we should examine the climatological time series used in the following analyses for variation in rates of warming across latitudes according to the period of observation. Here, a key focus of our investigation should be whether any increase in the rate of warming with latitude has been more apparent, or more pronounced, in more recent than in more historical segments of the period covered by the phenological data used in the preceding meta-analysis (1927–2013). To answer this question, we will examine high spatial resolution surface temperature data from the U.S. National Oceanic and Atmospheric Association (NOAA) Earth System Research Laboratory (ESRL; see appendix A).

Land-surface temperature anomalies for the annual growing season (April through June) are available from the ESRL at 0.5-degree latitudinal increments from 1900 through 2010. However, because we are concerned with phenological dynamics between 1927 and 2013, and between latitude 31.9 degrees N and 74.5 degrees N, the temperature anomaly time series will be similarly constrained in time and space. Determining latitudinal variation in rates of warming, and doing so in a temporally nonstatic approach, requires several steps. First, latitude-specific temperature anomalies for the northern hemisphere spring season (April through June) are regressed against "year" to derive estimates of linear trends in the anomalies (Post et al. 2018). For reference, the anomalies in the ESRL database are relative to the mean temperature for 1951 to 1980. Next, these trends ("year" coefficients from the preceding analysis) are regressed against latitude (degrees North). This was done for the entire period covered by the phenology data, and for successively one-decade shorter periods—that is, 1938–2010; 1948–2010; 1958–2010; 1968–2010; 1978–2010; 1988–2010; and 1998–2010 (Post et al. 2018). This time-windowing approach, employed commonly in spatiotemporal analyses of population dynamics (Ranta et al. 1997; Ranta 1998; Rodó et al. 2002), was adopted as a means of examining variation in latitudinal warming trends both with length of the period of observation and with proximity to the present because, as noted previously, shorter and more recent phenology time series displayed stronger trends than did longer and older time series (figures 2.8 and 2.10; Post et al. 2018).

Figure 2.13a depicts variation in rates of warming with latitude across these successively truncated and more recent periods. The periods with the greatest average rate of warming, calculated within the latitudinal time series, were the three most recent (1978–2010, 1988–2010, and 1998–2010), while that with the lowest rate of warming was 1928–2010 (mean rate of warming = 0.104°C per decade) (figure 2.13a; Post et al. 2018). Moreover, the greatest rates of warming occurred in the most recent period (1998–2010) above latitude 61.5 degrees N,

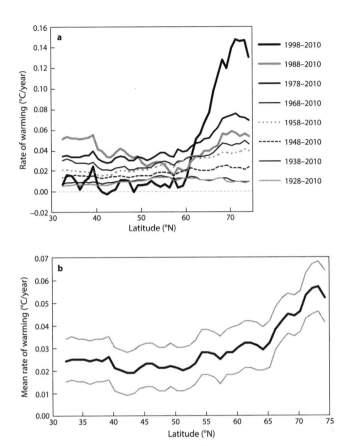

FIGURE 2.13. Panel (a) depicts variation in the association between latitude (degrees north) and rate of warming (degrees Celsius per year) over land. The latitudinal range depicted in each plot is 32.5 degrees N to 74.5 degrees N. At each latitude, the rate of warming was estimated using linear regression, with annual April–June (inclusive) temperature anomalies as the dependent variable, and year as the predictor variable for the periods defined in the panel legend. The resulting rate coefficients are plotted against latitude in one-degree increments to obtain the latitudinal series for each time period. Annual April–June temperature anomalies were obtained from the University of Delaware land-surface temperature data, which are available at 0.5-degree spatial resolution for the period 1900–2010. Annual anomalies are calculated based on the deviation from the mean for the period 1951–1980. The spatiotemporal range of the data used in this analysis was chosen to correspond most closely with that of the phenology data used in the meta-analysis presented in this chapter. Panel (b) depicts the increase in the mean rate of warming (black line), calculated across all periods in panel (a), with increasing latitude (degrees north). The latitudinal mean rate of warming was calculated using only significant ($P < 0.05$) year coefficients from the regression analyses described in the legend to panel (a). These were also used as the basis for estimating the rate of increase in the rate of warming with latitude for purposes of comparison to increases in the rate of phenological advance with latitude. The gray lines in panel (b) represent 95 percent confidence bands. Modified from Post et al. (2018).

approaching the Arctic zone (figure 2.13a; Post et al. 2018). Significant ($P < 0.05$) increases in the rate of warming with latitude were detected for all periods except 1988–2010, for which there was neither a positive nor a negative trend in warming with latitude (figure 2.13a). Quadratic functions improved the fit of the relationship between rate of warming and latitude for all periods except 1948–2010 and 1958–2010, suggesting that for all other periods, the rate of warming varied nonlinearly with latitude. This is most evident visually for the periods 1998–2010 and 1978–2010 (figure 2.13a; Post et al. 2018). We will return to this in a moment.

Next, it is necessary to estimate the latitudinal acceleration of warming, or the rate at which the rate of warming increases with latitude. Simultaneously, we must account for variation owing to length and start of the observational time series upon which warming rates were based. Preferably, this should be done in an approach that is comparable to the analytical approach used to account for variation in time series start year and latitude in estimating the rate of phenological advance with latitude. Hence, here again we can employ a generalized additive model that includes latitude as a numerical predictor and time series start year as an ordinal predictor. This analysis reveals that the rate of warming increases significantly with latitude (linear regression $P < 0.05$; GAM standardized regression coefficient = 0.34 ± 0.04, $P < 0.05$) (Post et al. 2018). Visually, we can see that the rate of warming at the highest latitudes represented in the phenology data set (from 66 degrees N to 74.5 degrees N) is approximately double that at lower latitudes (figure 2.13b; Post et al. 2018). However, to test more rigorously whether rates of warming vary with latitude and period of observation, it is likely necessary to employ piecewise nonlinear regression (Framstad et al. 1997; Stenseth et al. 1998a, 1998b) of the coefficients in figure 2.13a. This is because the greatest rates of warming were observed in the periods 1978–2010, 1988–2010, and 1998–2010, and because the relationship between rate of warming and latitude in each of these periods was best fit by a quadratic function.

In this approach, the relationship between rate of warming and latitude for any given period is analyzed according to the threshold function:

$$Y_i = \begin{cases} \alpha_{1,0} + \alpha_{1,1} X_i + \varepsilon_{1,i}, X < \Theta \\ \alpha_{2,0} + \alpha_{2,1} X_i + \varepsilon_{2,i}, X \geq \Theta \end{cases} \tag{2.1}$$

in which Y_i is the rate of warming at latitude i, X is latitude, and Θ defines the threshold in latitude above and below which the coefficient α is expected to differ. Applying such an approach, the greatest rate of increase in the rate of warming with latitude is revealed to occur during the most recent period, 1998–2010, and north of 59–61 degrees N latitude, when and where this association is nearly an order of magnitude greater than in the next greatest periods, 1988–2010 and 1978–2010, at the same range of latitudes (figure 2.14; Post et al. 2018). Hence,

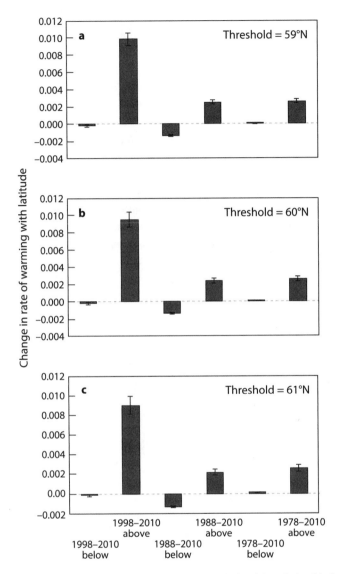

FIGURE 2.14. Results of piecewise, nonlinear regression analysis of the relationship between the rate of land-surface warming and latitude for the periods 1998–2010, 1988–2010, and 1978–2010, with a threshold criterion for latitude, according to equation (2.1). The threshold is set at 59 degrees N latitude (a), 60 degrees N latitude (b), and 61 degrees N latitude (c). These periods were selected for piecewise nonlinear regression because the relationship between rate of warming and latitude during each of them was best fit by a quadratic function. The x-axis categories define the coefficient estimates (±1 SE) below (that is, south of) and above (that is, north of) the indicated threshold latitude. Hence, the period 1998–2010 was characterized by the greatest rate of increase in the rate of land-surface warming with latitude north of 59–61 degrees N. From Post et al. (2018).

the pattern of greater rates of phenological advance at high compared to low latitudes is not simply an artifact of shorter and more recent time series that coincide with the onset of rapid warming. There is reason to accept that they represent responses to rapid warming.

The alleviation of abiotic environmental constraints on the timing of expression of life history events advances phenology more rapidly as one moves poleward, at least in the northern hemisphere. This pattern suggests, regardless of the mechanism through which alleviation of abiotic environmental constraints on phenology operates, such alleviation acts to increase the availability of time for biological activity on a seasonal basis. But how can the availability of time, which we might be tempted to think progresses with inexorable linearity and unidirectionality, be altered by changing environmental conditions? As the next chapter will argue, the answer may lie in defining the nature of time and its role as an ecological resource at the level of the individual organism.

CHAPTER THREE

Ecological Time

In ecology, time is traditionally considered a niche dimension along which organisms, usually of competing species, may segregate to minimize overlap in the use of other resources (Schoener 1974; Kronfeld-Schor and Dayan 2003). A classic example illustrating such segregation is that of differential partitioning of time on an annual and subannual basis by species of rodents that consume similar-size seeds in the Chihuahuan Desert (Brown and Munger 1985). Similarly, in a Mediterranean oak forest on Mount Holomontas, Greece, insects in an assemblage of 58 co-occurring species partition time on a seasonal basis and thereby reduce temporal overlap while foraging (Kalapanida and Petrakis 2012). Even finer scale partitioning of time is evident in the foraging activity of two congeneric species of gerbils in the Negev Desert, Israel, that segregate between early- and late-night foraging, thereby minimizing temporal competition for seeds (Ziv et al. 1993). Closely related species may also partition time on similarly fine scales to minimize competition for mutualists. A case study on the timing of pollen release by co-occurring species of *Acacia* sp. in Tanzania (Stone et al. 1996), presented in greater detail in chapter 7, provides a compelling example of this. The classic conceptualization of the temporal dimension as a niche axis also pertains to the partitioning of activity seasonally and during the 24-hour diel cycle as a predator-avoidance strategy (Lima and Dill 1990; Skelly 1994; Lima 1998). But such conceptualizations of time as a niche dimension appear to apply mainly, if not exclusively, in the context of behavior related to resource acquisition and use.

More recently, there has been a growing emphasis on recognition of time as one of two major axes—the other being space—across which ecological dynamics unfold and across which ecological systems are shaped (Pau et al. 2011; Wolkovich et al. 2014a, 2014b; Park 2017). The rationale underlying this argument draws many parallels, and important contrasts, between the roles of time and space in ecology. Chief among these are that ecological patterns in time, like those across space, may be autocorrelated; that variation across environmental gradients of time, like that across space, influences the distributions of organisms; and that time, unlike space, introduces unidirectionality to ecological patterns and processes (Wolkovich et al. 2014b).

These arguments are vital to an understanding of phenological dynamics. But applying them to develop an understanding of why the expression of life history traits related to the timing of events varies to such a considerable extent across species, within species, and even within organisms requires additional nuance. As the preceding chapter emphasized, while there are clear general patterns in phenological dynamics across levels of biological organization and across scales of space, there are important exceptions to these patterns. Some taxa show little or no phenological advance even when the presumed constraints on the expression of the timing of life history traits are alleviated. Likewise, taxa that do advance the timing of one or more life history events coincident with the alleviation of constraints on their expression do not advance the timing of other events. In some cases, taxa, or even individual organisms, delay the timing of some life history events in response to the same environmental changes that elicit an advance in the timing of other life history events.

To understand such variable patterns in phenological dynamics, and to explain them in an evolutionary context, we should explore the role of time in ecology not only as a niche dimension or an axis of interactions, but also as a resource. Yet ecology, as already alluded to, has had an admittedly troubled past with the concept of time. Callender's quote at the start of this book reminds us that physics is alone among the sciences in regarding time itself as worthy of formalized consideration. In ecology, time is a parameter, or occasionally an exponent or a subscript, in conceptualizations of other processes or patterns, or in equations describing them. And even within this arena, time is treated in an inconsistent fashion, sometimes as discrete and other times as continuous. Is this because ecologists are reluctant to grant time the status afforded it in physics, where time is a central focus from which many other aspects of the field emanate? If so, consideration of perspectives underlying the notion that space is a resource might foster an appreciation for the role of time as a resource.

When we consider space, we intuitively understand that it is in limited supply. In our daily lives, the use of space by an inanimate object or by someone else renders that space unavailable for use by another object or by ourselves. Likewise, in ecology, the use of space by one organism renders that space unavailable for use by other organisms of the same or other species. But when we consider the limited nature of space in such contexts, we operate under the assumption that time is constant. In other words, use of space by one organism renders it useless by another organism *only for the time being*. Use of space at time t does not, by necessity, render it useless by a potential competitor for that space at time $t+1$. The same, however, cannot be said for time. Time on ecological scales unfolds universally across the spatial dimension. And while the use of space may be applied to multiple purposes, once time is allocated to a transition from one life history stage to another it cannot subsequently be reallocated to a preceding life history stage.

This latter assertion requires some elaboration. Obviously, time can be, and is, applied to multiple processes simultaneously by every living organism. A plant may, for instance, continue to grow while flowering. And a bird invests in its own maintenance and survival while incubating its eggs on the nest. However, the transition from one phenophase to the next in an organism's life cycle cannot be reversed. Hence, the allocation of time to any given phenophase renders that time unavailable for allocation to another phenophase. A brief analogy may serve to further illustrate this point. Consider the process of allocating time to your own success. In academics, for instance, this likely involves investing time in processes or products that, in the case of students, contribute to performance metrics such as grades, and, in the case of researchers, contribute to outcomes such as funding. With multiple competing demands for time, a student must invest time in studying for one exam at the expense of studying for another, while a researcher must allocate time to preparing one proposal at the expense of preparing another or at the expense of writing a paper. And any such allocation or investment of time cannot be reversed for reinvestment in some other similar process or product. In a life history context, this illustrates a central challenge faced by all organisms: contending with the inexorable, unvarying, and unidirectional passage of cosmological time in the process of contributing successfully to future generations. Lost cosmological time cannot be easily regained, if at all. The recurrent and cyclical nature of diurnal and seasonal time, and the varying availability of relative ecological time, however, afford organisms the opportunity to contend with the passage of unidirectional cosmological time, as will be discussed later in this chapter.

TIME AS A RESOURCE IN ECOLOGY

Time is a prominent element of many facets of ecology and evolution. It can refer to the occurrence of an event, such as the classic components of phenology, the divergence of sympatric species, or extinction. In such cases, we use the term *timing*. It can also refer to the pace at which phenological events unfold within years or advance across years, or to the pace of speciation or extinction. In these cases, we use the term *rate*. And it can refer to the length of a particular phenological event or process in speciation or evolution, which is termed *duration*. Timing, rate, acceleration, and duration are the traditional contexts in which time is given consideration in phenology in particular and in ecology and evolution in general.

But to understand phenological dynamics in a life history context, it may very well be useful to consider time as a resource. In this framework, the established terms *timing*, *rate*, *acceleration*, and *duration* accrue more nuanced importance. Timing refers to the acquisition or capture of time, rate refers to the pace at which time is used or allocated, acceleration refers to a change in the rate of use or allocation of time, and duration quantifies the amount of time applied or allocated to any particular stage in the life history cycle. Here, it may be fruitful to briefly draw a distinction between the use of such terms in this context and the more familiar use of such terms in ecophysiology. In the latter, timing and duration refer to the occurrence and pace of progression of metabolic reactions and metabolic rates, which scale with organism size (Schmidt-Nielsen 1984), a relationship that underpins the metabolic theory of ecology (West et al. 1997; Enquist et al. 2003; Brown et al. 2004).

Here, in contrast, timing, rate, and duration refer to the onset of, progression through, and hence use of time in, linked series of associated phenophases that comprise an individual organism's life history cycle (Ehrlen 2015). As with other resources, time available for allocation to such phenophases or life history stages is limited in supply, and it is in demand by conspecific and heterospecific competitors alike. It also limits the availability of other resources needed for life cycle progression that are likewise in shared demand by competitors. It may even be suggested that time, the resource, drives the evolution of life history traits, variation in the expression of which is studied as phenology. After all, it could be argued that the evolution of clock genes, photoperiodic endocrine function, diurnal periodicity in patterns of behavior, and annual cycles of activity are compelling evidence of selection for a variety of means of time-keeping by the individual organism (Aschoff 1960, 1966; Aronson et al. 1993; van Oort et al. 2005, 2007; Lu et al. 2010).

As alluded to in the introduction, the intent here is to promote the notion of time as a resource. This argument necessitates reviewing the conditions that must be met for a state of competition to exist: there must be a shared requirement for a limited resource. Not only is time in limited supply, it is also absolutely required by all species and by all individuals of conspecific species. Moreover, time limits the availability, acquisition, and allocation of other resources to growth, maintenance, and reproduction. If we view life cycles in the context of the second law of thermodynamics, then time is a necessary component in the expenditure of energy in the processes of growth, maintenance, and offspring production, all of which oppose entropy. In fact, we might consider life itself the process of allocating time to gene propagation or the conversion of time to biomass.

Here we must ask ourselves whether time can justifiably be considered a *limited* resource. The answer lies in how we consider the use of time. The use of time by one organism does not necessarily alter its availability for use by another organism, so absolute cosmological time might not be, in this sense, a limited resource. But other forms of time, about which more will be explored later in this chapter, might be competed for among individuals. One such form will be referred to subsequently as *relative ecological time*, which is represented by the timing of activity of other organisms that are used as resources in all forms of organism-organism interactions. However, time, whether absolute cosmological time or relative ecological time, is most clearly in limited supply for the individual organism. Time, unlike other resources, cannot be stored. There is only so much time within an individual's existence to grow and reproduce. As well, the availability and use of time by the individual may influence the availability and use of other resources. For instance, suppose that in a plant community, nitrogen is in limited supply. Suppose as well that time is in limited supply within the same community—time for initiation of growth, time for development, and time for reproduction. Now, suppose from the perspective of the individual plant, of the two resources considered so far, time is more limited. Then, rather than viewing the individual's use of time as primarily devoted to partitioning the limited availability of nitrogen, we might realize the individual's use of nitrogen also facilitates its capture of time at the expense of competitors. We might therefore consider time to be a universal resource. Hence, the division of time, or the division of demand along the niche axis comprising time, is a primary component of niche complementarity. And the more similar the resource demands for any two organisms, the more intense will be their competition for that resource if it is in limited supply. Thus, individuals must balance competition with conspecifics, with which they have identical resource requirements, and heterospecifics, with which they have at least somewhat divergent resource requirements.

If time is indeed a resource, and in limited supply, then natural selection should favor the evolution of strategies by individuals to compete for and acquire more of it. But how, we might wonder, can an individual acquire more time? The simple answer is by becoming seasonally active earlier than others, by remaining seasonally active later than others, or both, thereby prolonging the duration of critical phenophases determining successful offspring production. Alternatively, individuals may partition time to minimize overlap along the temporal niche axis with competitors. If, however, time is a resource, and a limited one at that, then how do we explain the fact that some species advance their phenology, while others do not alter their phenology, and yet others delay theirs, in response to changes in biotic and abiotic environmental conditions? One proximal explanation is that other constraints on phenological advancement limit shifts in the timing of expression of life history traits.

While temperature and precipitation have received, by far, the greatest focus in studies of environmental drivers of phenological dynamics (Schwartz 2003), limitations on phenology imposed by photoperiod and solar irradiance are of clear importance in many species and systems as well. Notable examples with relevance to species' responses to climate change include seasonal cycles of inappetence and related thyroxin release in caribou and reindeer (Ryg and Jacobsen 1982; Tyler et al. 1999), the timing of seasonal pelage molt in snowshoe hares (Mills et al. 2013; Zimova et al. 2014), initiation of seasonal growth in some species of plants (Shaver and Billings 1977), and the initiation and pace of long-distance migration in some species of birds (Clausen and Clausen 2013). Hence, primary limitations on, or drivers of, the expression of life history traits related to timing include, among abiotic factors, temperature, precipitation or moisture, photoperiod, and solar irradiance; and, among biotic factors, interactions with other individuals, whether conspecific or heterospecific (Kronfeld-Schor and Dayan 2003; Johansson et al. 2015a, 2015b). Ascribing divergent phenological patterns, such as phenological stasis, advance, and delay, to the operation of such a variety of constraints does not, however, explain why or how sensitivity to a variety of such constraints has evolved. This point will be revisited in chapter 7.

The traditional conceptualization of time in ecology, then, is as an axis within the n-dimensional niche hypervolume along which species may segregate to minimize competition or interaction with natural enemies (Schoener 1974; Richards 2002; Kronfeld-Schor and Dayan 2003; Howerton and Mench 2014). In such a perspective, time is a dimension that facilitates utilization of other axes in the n-dimensional niche hypervolume in a manner that reduces competition or the risk of encountering competitors or natural enemies. The suggestion that time itself is a resource implies that time is not simply a dimension—like space—across which ecological interactions and dynamics unfold. Rather, time determines *how* ecological dynamics and interactions unfold, and *why* they do so in the manner in which they unfold.

Several key insights derived from spatial ecology can be applied to the study of phenology to assist in this conceptualization. For instance, one of the major challenges to the development of the field of spatial ecology concerns the conflation of the potential role of space itself in ecological interactions with the role of space as a mediator of the roles of other factors (Steinberg and Kareiva 1997; Tilman 1997). This applies as well to the role of time in ecological interactions. Moreover, as with space, time may be heterogeneous with respect to resource distribution and availability. Even across a temporally invariant landscape, the spatial distribution and availability of resources could be highly variable. Similarly, in a spatially invariant landscape, resource distribution and availability may be highly variable through time. Such variability would connote seasonality, which is, as discussed earlier, distinct from phenology.

As well, as with space, time may be littered with or devoid of intra- and interspecific competitors. Landscapes may represent a mosaic of hotspots of spatial overlap among conspecifics and heterospecific competitors independently of any temporal variation in their abundances, or may contrastingly provide refuges from them where their abundances are low. Similarly, the temporal environment in a spatially invariant landscape may become alternately crowded with conspecific or heterospecific competitors, or devoid of them as seasonal and longer-term fluctuations in abundance unfold. And last, as with space, time may harbor refuge from natural enemies, or provide opportunities for beneficial interactions with conspecifics, or with heterospecific individuals in mutualistic species interactions. Spatial heterogeneity can provide enemy-free zones across the landscape or zones of predictable overlap among the distributions of species dependent upon one another for mutualistic interactions. Similarly, in a spatially invariant landscape, temporal variation on seasonal or longer time scales may result in enemy-free periods or periods of convergence of the activities of species engaged in mutualisms (Jaksic 1982; Wheelwright 1985; Skelly 1994; Aizen and Rovere 2010). The role of time in lateral, interference species interactions and in vertical, exploitative species interactions, will be a major focus in chapters 6 and 7, respectively.

Admittedly, it is difficult to conceptualize time as a limited resource when considering competition among individuals because it is difficult to envisage the use of time by one organism interfering with its use by another organism. To address this challenge, let us consider competition in a slightly different manner than that implied thus far. Is there, for instance, a situation in which the use or allocation of time by an organism does in fact reduce or remove its availability for other uses or allocation? One such possibility involves the use or allocation of time within an individual. Hence, while the allocation of time by an individual plant in the process of development during a life history stage preceding reproduction, for example, may not interfere with the use of time during the same period by another individual plant, it does in fact interfere with the allocation of that same time by the individual to any other life history stage. The analogy presented earlier in this chapter to the allocation of time in studying for an exam or writing a proposal helps to reinforce this notion. Viewing phenology in such a life cycle context (Post et al. 2008a; Ehrlen 2015) highlights the temporal dependence among sequential life history stages. It also emphasizes, importantly, that the timing of expression of any given life history event relates both to environmental cues and limitations as well as to the timing of other life history events within the life cycle (Ehrlen 2015). Hence, time may differ from other resources in being limited in its availability to the individual organism. If so, competition for time occurs not only among individuals of the same or differing species but also within individuals themselves among life history stages.

If this is indeed the clearest scenario under which time may be conceived of as a limited resource, then where in the study of phenology should we look for examples of the allocation of time as a competitive strategy? Plasticity in the timing of some life history events and a lack of plasticity, or less of it, in the timing of other life history events, is consistent with the notion of time as a limited resource (Post et al. 2008a; Phillimore et al. 2012; Ehrlen 2015). Examples of this abound in the literature. For instance, some plant species respond to observed or experimental warming by advancing their timing of leaf-out or emergence, yet they do not comparably advance their timing of flowering and in some cases may even delay it (Galen and Stanton 1995; Parmesan 2007; Post et al. 2008a; Forrest and Miller-Rushing 2010). Some species of migratory birds differentially advance their timing of migration or arrival at breeding sites and their timing of nesting or egg-laying (Both and Visser 2001). Such patterns, at the very least, indicate that not all life history events in sequence are equally responsive to the alleviation of constraints on the timing of their expression. They also illustrate that the allocation of time to the expression of one life history event occurs at the expense of the allocation of time to the expression of other life history events. This is consistent with the notion of time as a limited resource.

SCALES OF TIME IN THE DOMAIN OF PHENOLOGY

Just as other resources may be available in multiple forms, time is available at multiple scales. For instance, nitrogen is available in elemental form, or in multiple compound forms such as amino acids, ammonia, and nitrate. Time is available for use by the individual organism as milliseconds, seconds, minutes, hours, days, weeks, or even years and decades. These are, to great extent, related to metabolic rates and body size or mass of the organism (Schmidt-Nielsen 1984; West et al. 1997; Brown et al. 2004), and, to some extent, defined by the scales at which time is quantified according to convenient units of measure, some of which are arbitrary. But they are also defined in absolute and relative terms.

We have already touched briefly upon applications of time in ecophysiology and metabolic scaling laws. Additionally, the term *ecological time* has been applied previously in the context of phenology, and distinguished from clock time and physiological time, as the "interval during which a defined and constant number of demographic events (i.e, births plus deaths) occurs" (Stamou et al. 1993). In the context of the availability of time as a resource, there are at least three forms or types of time. First, there is cosmological time, which is absolute, and commonly referred to as clock time. Second, there is cyclical or recurrent time, which is determined by the rotation and tilt of the Earth, as well as its orbit around the Sun. This is the form of time that determines diel light:dark ratios, the seasons, and the passage of years. Third, there is relative ecological time. This refers to the timing of biological events and ecological interactions relative to other biological events and ecological interactions. While an insect pollinator may emerge early in an absolute sense one spring in comparison to a long-term average of emergence dates, such timing may be late relative to the timing of flowering that same year by a species of plant it pollinates. Similarly, while the duration of flowering by that plant species may be reduced in absolute duration one summer in comparison to a long-term mean, the duration of flowering may be relatively extended that season from the perspective of its pollinator if flowering and pollinator presence overlap for longer that year than in other years. Hence, the absolute timing of the annual period of growth, development, or reproductive activity of an organism may be described in terms of seasonality, as in "early or late spring." The relative timing of the same activity may be described in terms of earlier or later than that of conspecifics in the same population or co-occurring species in the same local species assemblage. We will return to distinctions between cosmological time, recurrent time, and relative ecological time later in this chapter.

Scales of time in an ecological context are also inextricably intertwined with scales of space (figure 3.1). Short-term processes unfold at subcellular, cellular, and organismal scales, but not at superorganismal scales. For instance, metabolic reactions, the Krebs cycle, and cell division related to mitosis for growth and maintenance and meiosis related to reproduction all occur at the shortest time scales, from milliseconds to seconds (figure 3.1). The expression of life history traits related to timing, including those concerned with growth, productivity, and offspring production, unfold over scales of time between the horal and the annual, and at scales of space limited by the individual organism. Interactions among species occur over horal, diel, and annual time scales, and occur at the spatial scale of individual organisms. Long-term processes, on the other hand, unfold over large scales, but not at organismal or suborganismal scales. These may include, for example, population dynamics, the formation and dissolution of species assemblages, succession, and speciation, all of which occur over spatial scales that are superorganismal and extend to the landscape and regional scales (figure 3.1).

What, then, are the relevant time scales for individual-level processes related to survival and reproduction? They are the temporal scales over which and within which the individual organism's life history cycle unfolds. And these are the same scales that encompass phenology and its study and measure in plants and animals: days, weeks, months, and years. These are the scales of time in figure 3.1 at and below the annual scale. Units of time longer than these, and in many cases at the upper end of this range of variation (for example, centuries and millennia to supramillennia), are unavailable to the individual organism for incorporation into the timing of life history events and development of the life history cycle. These are the scales of time that exceed the annual scale in figure 3.1, although the

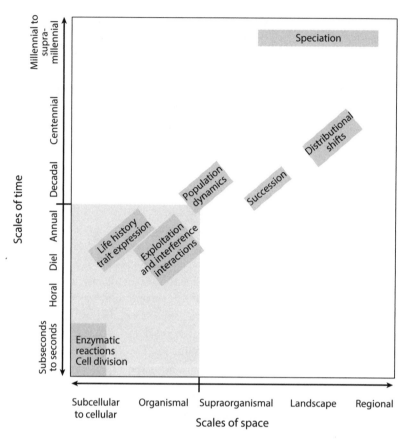

FIGURE 3.1. Relation between the scales of time over which biological, ecological, and evolutionary processes occur and the scales of space over which they operate. Those processes occurring at the shortest time scales—subseconds to seconds—tend to occur at the suborganismal scale. Above this temporal range, scales of time available to the individual organism for allocation to activity, growth and development, and reproduction extend to the annual scale. Processes unfolding at time scales beyond the annual tend to represent supraorganismal dynamics, including population and community dynamics, distributional shifts, and speciation. Hence, time accessible by the individual organism, which comprises resource time, tends to fall within the scales indicated by the gray shaded area, from the annual scale and below.

interval between annual and decadal may also constitute scales of time available to the individual organism in long-lived species. Hence, the range of space-time scales bounded by the individual organism along the spatial axis and bounded by the annual period on the temporal axis represent the region of space-time relevant in the domain of phenology. The justification for delimiting ecological time in this context to that unfolding at the annual scale or below will be elaborated upon later. For now, however, we may be left wondering how the individual organism acquires time and incorporates it into a life history cycle such as the idealized cycles depicted in the next section. The answer lies in drawing a distinction between the timing and duration of life history events.

TIMING AND DURATION OF LIFE HISTORY EVENTS

By far, the bulk of phenological studies reviewed here have focused on the timing of individual life history events, and most of these relate to early-season events. But to understand why patterns of variation in the timing of life history events can appear inconsistent within traits among species and across traits within species, we should consider phenology in the context of the adaptive value of life history traits to the individual organism. This thinking leads us to an understanding of the role of timing in the phenomenon of duration. What does an organism achieve by altering the timing of expression of one or more of its life history traits in response to, for example, the alleviation of environmental constraints? And what, furthermore, does it achieve by not adjusting the timing of expression of one or more life history traits? Advanced and more rapid progression through early life history stages and delayed progression through later life history stages can both achieve the same outcome: increased duration of some other life history stage that is perhaps more critical to growth, survival, and reproduction.

To illustrate the relationship between timing and duration, and the manner in which timing is, in essence, critical to duration, let us first examine three simplified life history schedules representing a univoltine invertebrate organism (figure 3.2a), a migratory vertebrate such as a bird or mammal (figure 3.2b), and a vascular plant (figure 3.2c). During an annual cycle, the timing of each individual event in the life history schedule of any of these idealized organisms represents the expression of a life history strategy.

The life history strategy in such an annual cycle reflects selection in response to multiple factors for the optimal expression of the timing of the individual life history events that the strategy as a whole comprises. The timing of expression of life history events that occur early in the annual cycle, such as emergence in invertebrates and plants, or spring migration and arrival at breeding grounds in migratory vertebrates, is under selection to coincide with the alleviation of environmental constraints on resource availability and thereby maximize resource access. Selection pressures in this case relate to both abiotic conditions and to biotic conditions, including interactions with conspecifics and heterospecifics. Hence, the optimal timing for expression of early-season life history traits may be as early as favorable abiotic conditions allow and before the biotic (competitive)

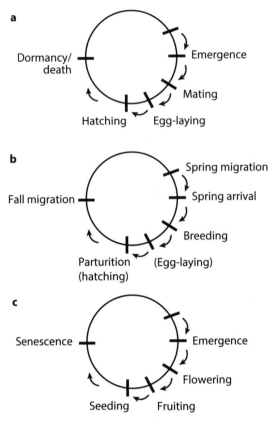

FIGURE 3.2. Idealized life history cycles of three hypothetical organisms: an arthropod (a), a migratory bird (b), and a perennial plant (c). The circles in each represent the annual cycle, while hatch marks indicate the relative timing of individual life history events, or phenophases, within the cycle. Autumn and spring events, synchronized with the autumnal and vernal equinoxes, respectively, are indicated by horizontal hatch marks at the left and right sides of each organism's cycle. For the individual organism, the allocation of time to any single life history event or phase defined by the timing of its onset renders that time unavailable for allocation to any other event. Hence, the limited nature of time as a resource applies most clearly in the context of its capture and allocation by the individual organism to events within the life history cycle.

environment becomes too crowded (Iwasa and Levin 1995), but not so early or late as to miss opportunities for overlap with resource species whose phenology also varies.

Importantly, the timing of early-season life history events sets the stage for the timing of successive events, each of which is increasingly under selective pressure owing to biotic factors that now include, in addition to density-dependent and interspecific competition, exposure to natural enemies (Post 2013; Wolkovich et al. 2014a). These selective pressures relate to the durations of the intervals between successive life history events, which are critical to meeting the costs of maintenance, growth, and development, all of which contribute to survival. Hence, the *timing* of individual life history events can be viewed as directed primarily toward determining the *duration* of critical stages between them. And owing to the unidirectional nature of time as an ecological agent (*sensu* Wolkovich et al. 2014b), time, once acquired and allocated as a resource to any single life history event, is rendered unavailable for allocation to successive life history events. Hence, the acquisition of time most clearly presents a selective advantage when allocated in the context of the duration of life history events.

THE ADAPTIVE VALUE OF PHENOLOGICAL ADVANCE, STASIS, AND DELAY

While it is perhaps common to interpret phenological advance as adaptive and stasis or delay as maladaptive, there may be conditions under which advance is maladaptive and stasis and delay are adaptive. The adaptive or maladaptive nature of phenological shifts or stasis will be determined primarily by two outcomes. The first of these is their effect on the duration of phenophases that are critical to survival and offspring production. This will be the focus of the current section. The second of these is their effect on the extent of overlap in time for species interactions that are either beneficial (for example, mutualistic or resource acquiring) or detrimental (for example, exploitative or interference). These latter effects will be the focus of chapters 4, 6, and 7.

To illustrate the manner in which the timing of life history events determines their duration, let us begin by unfolding the circular life history diagrams in figure 3.2. We can then lay out a segment of any of these life history cycles along a linear axis that represents the timing of the series of life history events comprising the annual life history cycle. For simplicity, let us consider the timing of two successive life history events in relation to one another (figure 3.3a). The changes in duration that derive from shifts in the timing of successive life history events apply to any sequence of events regardless of the number of events in the full life history cycle as long as we consider shifts in timing of events relative only to the immediately preceding or successive event.

The nine scenarios depicted in figure 3.3a illustrate the manner in which shifts in the timing of successive life history events in relation to one another can alter the duration of the interval between the events, and, hence, the duration of the first of the two life history events in the sequence. Throughout the discussion of the following scenarios, the described shifts in the timing of life history events refer to changes in the timing of a given event from one life history cycle to the next, or from one time step to the next. For instance, if we consider two successive life history events such as the annual timing of arrival at the breeding grounds by a migratory bird and the timing of initiation of nesting, then a shift in the timing of arrival refers to a change in its timing from year t to year $t + 1$.

The first three scenarios in figure 3.3a illustrate the three possible kinds of shifts in the timing of life history events that result in an increase in the duration of the interval between them, and hence an increase in the duration of the first of the two events in the sequence. Such an increase in duration may derive from an advance in the timing of expression of the first of the two successive life history

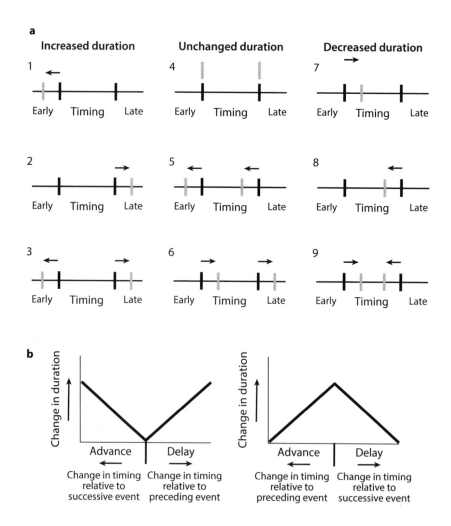

events, but no change in the timing of the second (scenario 1, figure 3.3a). Alternatively, an increase in duration may result from a *delay* in the timing of expression of the second of the two successive life history events and a lack of change in the timing of the preceding life history event (scenario 2, figure 3.3a). Last, an increase in duration may result from a simultaneous advance in the timing of the first life history event and a delay in the timing of the second life history event (scenario 3, figure 3.3a).

The next three scenarios in figure 3.3a illustrate the three possible ways in which the duration of the interval between successive life history events, or the duration of the first of two events in sequence, may remain unchanged. The first, and most obvious, of these, is a lack of shift in the timing of both the first and second life history events in the series (scenario 4, figure 3.3a). However, a lack of change in duration may also result despite shifts in the timing of life history events if the timing of both of the events in series advance to the same extent (scenario 5, figure 3.3a) or are delayed to the same extent (scenario 6, figure 3.3a) relative to one another. This form of dynamic should be characteristic of phenophases under selection for relatively fixed duration but variable timing, such as gestation length or incubation in vertebrates.

The final three scenarios illustrate the manner in which the duration of the interval between successive events, or the duration of the first of two events in sequence, may be reduced by shifts in the timing of these events. First, a reduction

---

FIGURE 3.3. (a) A representation of the three possible outcomes of shifts in the timing of life history events in a two-event sequence. First, a shift in the timing of one or both events in the sequence may increase the duration of the *phenophase*, defined as the interval between successive events. An increase in duration of the phenophase may occur if the first event advances while the second event remains unchanged (subpanel 1); if the first event remains fixed while the second event becomes delayed (subpanel 2); or if the first event advances while the second event becomes delayed (subpanel 3). The duration of the phenophase may remain unchanged if neither event shifts (subpanel 4); if both events advance equally (subpanel 5); or if both events become delayed equally (subpanel 6). Last, the duration of the phenophase may be reduced if the first event is delayed while the second event remains unchanged (subpanel 7); if the first event remains fixed while the second event advances (subpanel 8); or if the first event is delayed and the second event advances (subpanel 9). Each of these scenarios, arising from all possible combinations of advances, delays, or stasis, apply as well to the dynamics and duration of any interevent phenophase in the individual's life cycle, regardless of the number of events in the cycle. Hence, as indicated in panel (b), both advances and delays in the timing of individual life history events may result in increases or decreases in the duration of phenophases or intervals. In this context, therefore, both types of shifts in phenology—advances and delays—can be viewed as means to the same end: altering the duration of phenophases critical to growth or development, survival, and reproduction.

in duration may result from a delay in the timing of the first event and a lack of change in the timing of the successive event (scenario 7, figure 3.3a). Conversely, an advance in the timing of the second event in series concomitantly with a lack of change in the timing of the preceding event will also result in a decreased duration of the interval between the two events and of the first event itself (scenario 8, figure 3.3a). Last, a reduction in the duration of the interval between two events, and of the first event itself, will result from a delay in the timing of the first event and an advance in the timing of the successive event (scenario 9, figure 3.3b).

From the preceding scenarios, we may develop the following general conclusions. First, phenological advances, phenological stasis, and phenological delays may all result in increases in the duration of the interval between successive life history events and the duration of the first of two sequential life history events itself (figure 3.3b, left panel). We may view this as a strategy for the increased acquisition of or allocation of time to some critical life history stage, the duration of which is determined by the timing of the two life history events that bound it on either end. Generally, phenological advances, stasis, and delays are treated as disparate types of responses to the alleviation or intensification of abiotic constraints on the expression of life history events concerned with timing. In the framework presented here, they are seen as means to the same end in terms of the life history strategies of individual organisms. Each of them may represent a strategy relating to the use of time. This insight reveals itself, however, only when the timing of individual life history events is viewed in the context of a series of life history events wherein altering the *duration* of individual events or intervals between them is the strategy that drives shifts in the timing of successive events.

Second, neither advances nor delays in the timing of life history events result by necessity in any change at all in the duration of the interval between successive events. Nor do they result necessarily in increases in the duration of such intervals. In fact, both phenological advances and delays can *reduce* the duration of the interval between successive events (figure 3.3b, right panel) or the duration of individual life history stages. From an evolutionary perspective, it may be adaptive under some conditions to progress as rapidly as possible through a given life history stage, particularly if that stage is vulnerable to exploitation by natural enemies (Williams 1966). In the context of a theoretical framework of time in ecology, the prioritization of the acquisition and allocation by the individual of resources other than time, such as nitrogen or carbon allocated to growth, would represent an adaptive strategy in this context. As well, rapid progression through one particular life history stage may constitute an adaptive strategy aimed at allocating more time to either a preceding or subsequent life history stage. Such a strategy may be conceptualized rather easily by considering shifts in the timing of all individual life history events in the full sequence of the annual life history cycles in figure 3.2, rather than the simplified treatment of two-event sequences represented in figure 3.3a.

PARALLELS AND CONTRASTS BETWEEN THE NOTIONS OF SPACE AND TIME AS RESOURCES

In ecology, we can readily conceive of, envisage, and even measure the role of space in the dynamics of and interactions among organisms. Because of this, we can also readily conceive of space as a resource. The perception of space as a resource derives from the fact that, quite independently of scale, when we observe a system, we comprehend almost intuitively that the occupation of space by one organism or by a collection of organisms, precludes its use by other organisms. This applies as well to the suborganismal scale. The use of space by one part of a plant, for instance, precludes use of that same space by other parts of the same plant.

It is much more difficult, however, for us to perceive of time in the same way. In part, this relates to the fact that time, as noted previously, unfolds simultaneously across space, at least at scales relevant in ecology. But it is also owing in part to our inability to see time and to our consequent inability to perceive the utilization of time by living organisms. Time, it seems, is almost a background process in ecology and evolution, a descriptor of the unfolding of interactions and patterns, rather than a driver of them.

If we are willing to accept the concept of space-time—that is, the interdependence of space and time—and if, furthermore, we are willing to accept the notion that space is a resource, then we must also accept that time is a resource. If we are wiling to take this leap, then further aspects of the concept of space as a resource should apply to the concept of time as a resource. Let us explore some of these.

Space is not static through time. The surface area of space habitable by a given species may fluctuate through time, alternately becoming more or less abundant. For instance, falling and rising sea levels, retreating ice sheets, or the appearance of islands owing to volcanism are examples of processes that, over geological time scales, result in alteration of the amount of habitable space for terrestrial organisms. Similarly, climatic warming might open new habitat to a species while simultaneously rendering previously suitable areas occupied by it unsuitable. Range expansions and contractions poleward and upward in elevation illustrate this process. But does variation through time in the extent of habitable area for organisms truly connote variation in *space* through time? Not entirely, because it can be argued that such newly opened space always existed, on ecological time scales, or that recently closed space did as well. What changed was *access* to that space, or its availability.

But does the inverse apply to time? Is time variable through space? Yes, to some extent. The time that is usable for biological activity does in fact vary across space owing to the seasonality determined by the Earth's angle to the Sun and by its orbit around the Sun. This is the latitudinal gradient in seasonality described in chapter 2. And if this seasonality is altered by, for instance, climate change, then the amount of time usable by organisms in any given location may also be altered. Just as space may become available for colonization, time may become available for colonization as well. And, just as that space existed previously, but was unavailable for use, so too did that time exist previously, but access to it was obstructed by its use by other organisms or by other environmental constraints on access to it.

THREE FORMS OF TIME: COSMOLOGICAL, RECURRENT, AND RELATIVE ECOLOGICAL TIME

We may argue that, unlike the use of space, the use of time cannot be repeated. A mobile organism may make use of the same space repeatedly throughout time. The concepts of home ranges and territories, in which an organism repeatedly revisits the same locations within a subset of space, illustrate this principle. As reviewed in chapter 1, according to the cosmological theory of time, absolute time progresses unidirectionally. This conceptualization of time is commonly referred to as the "arrow of time" (Hawking 1985). Owing to the unidirectional flow of cosmological or absolute time, there is no temporal analogue in ecology for the repeated use of space by an organism through time because there is no such thing as the "same time," and so it can never recur. Time here refers not to the Newtonian notion of absolute time, which asserts that time is identical everywhere regardless of space (Newton 1687); rather it refers to the absolute passage of time in an ecological context as described earlier.

However, it is possible that an organism can reuse recurrent and relative time. In ecological time, in contrast to the unidirectional arrow of cosmological (or absolute) time, the notion of recurrent time relates to its cyclical nature (figure 3.4a). While absolute time progresses constantly forward, the same unit of time, no matter how finely or coarsely it is partitioned, never recurs. No existing organism has the capacity to reuse or reallocate any fraction of absolute time once it has passed. By contrast, cyclical time recurs with the diurnal, seasonal, and annual cycles that are superimposed upon the arrow of cosmological time owing to the Earth's rotation, its axial tilt, and its orbit around the Sun. Recurrent time,

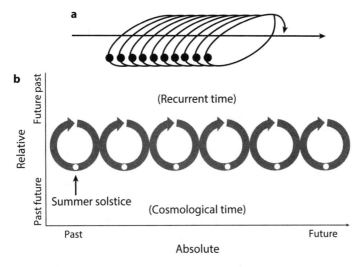

FIGURE 3.4. (a) Cosmological time, represented here by the linear "arrow of time," is absolute and progresses steadily from past to future. In cosmological time, no event is ever repeated, and individual events lie in either the past, the present, or the future, defined by discrete dates. In contrast, recurrent time is cyclic, and its cyclical nature as well as its scales are determined by the tilt of the Earth on its axis, the Earth's rotation about its axis, and the Earth's orbit of the Sun. Hence, recurrent time revisits the same events repeatedly. For instance, the summer solstice in the northern hemisphere, represented by the white dots, and the autumnal equinox, represented by the black dots, occur annually, albeit on different dates each year along the cosmological axis of time. In ecological and evolutionary contexts, the individual organism cannot recapture or reallocate cosmological time once it has been allocated or once it has passed without being allocated to growth, development, survival, or reproduction. In contrast, recurrent time may be reallocated or recaptured in the next or succeeding future cycles. It is the progression through cosmological time that allows the individual organism access to repeated units of recurrent time. (b) In this conceptualization, cosmological time progresses linearly along the x-axis, and comprises a constant relationship between past and future. The absolute nature of cosmological time dictates that its progression is unidirectional. Additionally, we see recurrent time arrayed as a series of repeating cycles that progress one into the next along the arrow of cosmological time. Hence, past and future in ecological time are not absolute, but rather are relative to one another. The relative nature of past and future in recurrent time can be visualized in this conceptualization, with the white dots representing the annual summer solstice and the gray dots representing the annual winter solstice in the northern hemisphere. The dates of each solstice comprise two elements: the year on which each falls, which varies along the x-axis with the passage of cosmological time; and the relative timing of each within any given year, which remains constant along the y-axis. Moreover, the summer solstice in any given year may constitute the relative past of the winter solstice in that year but the relative future of the winter solstice in the previous year, even though the timing of the summer solstice along the y-axis has not itself varied.

therefore, may be conceptualized as a spiral of time superimposed upon and progressing along with the cosmological arrow of time (figure 3.4a). The recurrent nature of cyclical time—and its contrast with linear and unidirectional cosmological time—becomes apparent in this conceptualization. Herein, we can see that while neither the 21st day of June nor the 23rd day of September in the year 2014 will ever recur, the summer solstice and the autumnal equinox in the northern hemisphere recur annually (figure 3.4a). While these dates have no inherent significance in the context of cosmological or absolute time, they have considerable importance in the context of recurrent time related to endocrine signaling that controls the expression of life history traits related to the timing of annual periods of production, reproduction, and migration.

To further elucidate the contrast between the linear and unidirectional nature of cosmological time and the cyclic nature of recurrent time, we may unfold the coils of cyclical time and plot them as an array of annual cycles along the cosmological arrow of time (figure 3.4b). Doing so facilitates recognition of the notion that time does not necessarily comprise a single axis. Rather, it may be decomposed into at least two axes. In addition to the linear axis that constitutes absolute cosmological time, there is the nonlinear axis that constitutes recurrent time. While cosmological time, according to the dynamic theory (see chapter 1), proceeds from past, through present, toward future, recurrent time, being cyclical in nature, oscillates between relative past and relative future, repeatedly revisiting the same relative dates, such as equinoxes and solstices. In this scenario, there is no absolute past and no absolute future in ecological time. Rather, the vernal equinox may function as a future event that will trigger springtime phenological activity such as migration timing, as well as a past event that establishes the annual starting point in the subsequent progression through a series of life history events that proceed toward a future summer solstice. This solstice, itself, becomes in turn a past event that determines the rate of phenological progression toward the next vernal equinox. According to this conceptualization, units of time with clear ecological meaning in a life history context would appear to be those below the decadal scale, while units of time at the decadal scale and beyond (that is, centuries, millennia, and so on) pertain less clearly to the expression of life history traits related to timing.

The context in which the notion of relative ecological time applies is less directly linked to cosmological time or to the tilt and motion of the Earth. In this context, relative time in ecology represents the timing of activity in other organisms with which the focal individual interacts. Such interactions may relate to interference or competitive interactions, to consumer-resource interactions, or to mutualisms such as host-parasitoid or plant-pollinator interactions. In this sense, the concept of relative ecological time refers to the timing of expression of life history traits relative to the timing of their expression in organisms with which interactions occur. These might include, as described earlier in this chapter, the timing of flowering by a plant relative to the timing of seasonal activity by its insect pollinator. Or, as will be explored in chapter 6, the timing of seasonal growth initiation by one plant species relative to its timing by other plant species with which it competes for resources in addition to time. As well, relative ecological time may be represented by the seasonal timing of initiation of plant growth in relation to the timing of seasonal arrival of migratory herbivores, a focus of chapter 7. Because the timing of such interactions can recur on multiple time scales both within seasons and years and among years, relative ecological time in this sense is cyclic as well. Hence, mobile organisms can make repeated use, in this context, of the same *relative* time through space, with relative time defined here according to the timing of activity by other organisms. For instance, use of *space* may enable a consumer to make repeated use of a resource at its peak of availability in *time* at many locations.

This brings us to another distinction between space and time as resources. An organism may move through time without moving through space. Sessile organisms, such as rooted plants, experience a very static spatial environment but a very dynamic temporal one. Such organisms do indeed "move" through time, experiencing its procession, without moving through space. Mobile organisms, on the other hand, are capable of moving through both time and space simultaneously, but they are also capable of moving through time while remaining fixed in space. Here, obviously, we are referring to relative stasis in space because the Earth itself is in motion. But no organism is capable of moving through space while remaining static in time. It follows, therefore, that organisms should be capable of utilizing time largely independently of space, but not space independently of time.

THE SPACE-TIME CONTINUUM IN A LIFE HISTORY CONTEXT

In a highly idealized setting, species can be arrayed along a spectrum of life history strategies concerning competitiveness between specialists that are superior competitors for space and those that are superior competitors for time. Hence, individuals of those species that are inferior competitors for space may be superior competitors for time, and vice versa.

Furthermore, organisms are capable of partitioning space, and of specializing on the utilization of space along a continuum from fine-grained to coarse-grained or from small scales to large scales. For instance, some species, most notably sessile organisms, specialize in the use of fixed space or on the use of small or confined spatial domains. Conversely, some species specialize in the use of large spatial domains. Even among sessile organisms, some species, such as large-canopy trees or clonal plants, occupy extensive spatial footprints. Among mobile organisms, nonmigratory species may occupy and utilize large territories or home ranges, while migratory species utilize seasonal ranges separated by large distances.

Possibly the most extreme example of the latter is represented by the annual migratory pattern of the Arctic tern (*Sterna paradisaea*), which involves migration from breeding sites in the Arctic to nonbreeding sites in the Antarctic, an annual round-trip migration of approximately 80,000 kilometers (Egevang et al. 2010). The northbound leg of this migration requires approximately 40 days, while the southbound segment requires approximately 93 days, on average (Egevang et al. 2010). If organisms are capable not only of partitioning space, but also of specializing, in their life history strategies, on the utilization of an array of spatial scales from local to extensive, might the same capacity for partitioning also apply to species' use of time?

The Arctic tern example illustrates the extent to which utilization of space is intertwined with the utilization of time. In general, we might expect long-distance migratory species, for instance, to require the use of more in that portion of their life history cycles devoted to migration than that used by short-distance migrant species of closely related taxa. Similarly, we might expect nonmobile species that specialize on the use of large spatial scales to specialize on the use of larger temporal scales than those used by species that partition space more finely. The phenological dynamics of some species indicate the capacity for finely resolved temporal plasticity. The meta-analyses reviewed in chapter 2 revealed considerable variation among taxa in their rates of recent phenological advance. Some species possess the capacity for rapid adjustment of the timing of expression of life history traits and hence advance their phenologies quickly in response to, for example, climatic warming or changes in the biotic environment. Other species appear more conservative, and advance their phenologies only slowly or not at all.

FRAGMENTATION OF TIME

Spatial ecology in both theory and practice has demonstrated that fragmentation of habitats increases the risk of extinction (Tilman and Lehman 1997). It does so by creating inhabitable zones within the distribution of a species, leading to distributional fragmentation and isolation of populations, reducing gene flow and exchange of individuals among them. This is a logical extension of the theory of island biogeography, in which it has been shown large islands proximal to mainland sources tend to harbor greater species richness than smaller islands distant from mainland sources (MacArthur and Wilson 1967). A principle analogous to spatial habitat fragmentation applies to the notion of the fragmentation of time. Key to the development of this analogy is the realization that, just as space still exists even after habitat fragmentation, it no longer represents a resource in the form of usable space. Similarly, time still exists after fragmentation, but it may no longer be usable by the organism of concern.

False springs occur when abiotic constraints such as temperature limitation or snow cover are alleviated through anomalously early warming in seasonal environments, followed by a return to cold temperatures or snow cover that persists until stable warming ensues (Marino et al. 2011; Peterson and Abatzoglou 2014; Allstadt et al. 2015). Under such conditions, organisms with an early life history strategy and that are in situ at the location of the early warming event may initiate activity in an attempt to maximize the acquisition of time for allocation to growth and reproduction in advance of potential competitors. Examples of the risks and adverse consequences of this strategy when cold or snowy conditions return are prevalent in the literature. For instance, early flowering forbs in the Rocky Mountains of Colorado, USA, suffer the loss of flowers, and hence opportunities for pollination by later emerging insect visitors, if early flowering is interrupted by spring freezes or snowstorms (Inouye 2008). Such losses can ultimately manifest at the community level, as shifts in species composition of local assemblages over the longer term if the reproductive success of individuals of such species declines as a result (CaraDonna et al. 2014).

Hence, the occurrence of false springs represents a compelling example of the manner in which the normally predictable pace of the progression of time from one season to the next can become interrupted. In consequence, the availability of time as a resource for allocation to successive life history stages in an annual life cycle can become fragmented through climate change. Ecological time may also become fragmented when biotic interactions are disrupted, such as in systems in which flowering is interrupted and then resumes within a single season (Crimmins et al. 2013). Such an interruption represents a break in the flowering stage of the idealized life history cycle presented in figure 3.2 for plants. Multiple flowering events within individuals or within species occur when flowering ceases without progressing to the fruiting stage, and then is reinitiated, after which the life cycle progresses. The obverse of this phenomenon—that is, uninterrupted flowering in a single event across a season—has been termed flowering constancy (Crimmins et al. 2013).

Instances of multiple flowering in this discontinuous, fragmentary fashion have been reported in systems with highly variable seasonal rainfall, such as tropical dry forests (Frankie et al. 1974; Opler et al. 1980; Bullock and Solismagallanes 1990), most commonly as occurrences of two episodes of flowering. Such phenological interruption likely constitutes a strategy, particularly in dry systems, related to drought tolerance (Crimmins et al. 2014). In the arid Sky Island plant communities of the southwestern United States, multiple flowering events are reportedly common among biennial species, herbaceous perennial species, and woody perennial species (Crimmins et al. 2014). However, this phenomenon is less common among annual species, which tend to display a strategy of early and rapid flowering following alleviation of arid conditions by episodic rainfall (Crimmins et al. 2014). Such interruptions of seasonal flowering, while presumably adaptive for the species engaging in this phenomenon, represent a clear response to fragmentation of relative ecological time. As well, they represent the fragmentation of resource phenology for consumer species such as pollinators, a topic that will receive more extensive treatment in chapter 7.

Having presented the arguments that time, much like space, varies in its scales of availability for use by the individual organism, and in its forms (absolute, recurrent, and relative) available for use by the individual organism, let us next consider how the use of time determines the manner in which the individual organism makes a living. To do so, we will develop the concept of the phenological niche.

CHAPTER FOUR

The Phenological Niche

Having examined the differences and parallels among the forms of time and the scales over which they unfold, and the extent to which they are accessible by living organisms, this chapter develops the concept of the phenological niche. This exposition will distinguish between describing the manner in which organisms use time in the expression of life history traits, in terms of absolutely early or late life history strategies, and the manner in which organisms use time in the expression of life history traits relative to their timing of expression by other organisms. This distinction will foster an understanding of how the timing of expression of life history traits and the duration of life history stages bounded by them result in strategies for the use of absolute time and relative ecological time. Development of the phenological niche concept will establish the framework for the phenological community and an examination of the role of time in the phenology of species interactions in the ensuing chapters.

THE ABSOLUTE PHENOLOGICAL NICHE

The concept of the absolute phenological niche originates from the notion that time comprises one among a multitude of axes defining the n-dimensional niche hypervolume (Kashkarov and Kurbatov 1930; Raney and Lachner 1946; Hutchinson 1957; Root 1967; Kronfeld-Schor and Dayan 2003; Elith and Leathwick 2009). According to conceptual models of what will be referred to here as the *absolute phenological niche*, species occupy time with varying degrees of specialization. For instance, within the same local assemblage, some species may exhibit active reproductive phenology year-round, while other species may exhibit nonoverlapping seasonal reproductive phenology, as is the case with flowering plants in the Wadi Degla desert ecosystem of Egypt (Hegazy et al. 2012). Hence, the individual life history strategies of species, or the seasonal patterns of species' individualistic expression of life history events, can be arrayed along a temporal axis (figure 4.1; Wolkovich and Cleland 2011). Along this axis, species may be classified according to their life history strategies as *early*, *middle*, or *late* species in absolute time.

Accordingly, any alteration of the availability of absolute time may have implications for biodiversity through effects of species packing or species loss, in the case of reduced availability of absolute time, or species addition through invasion of vacant niche space, in the case of increased availability of absolute time. Reductions of or increases in the availability of absolute time may result through an increase in the magnitude of abiotic environmental constraints, or through an alleviation of abiotic environmental constraints, respectively, on the onset or duration of the annual period of biological activity (Wolkovich and Cleland 2011, 2014).

Wolkovich and Cleland (2011) proposed four scenarios through which biodiversity may be altered by exotic species invasion of phenological niches: vacant niche effects, priority effects, niche breadth effects, and phenological plasticity effects. To this list may be added latency effects. Let us examine each of these in detail.

The occurrence of vacancy along the temporal niche axis may arise from short-term fluctuation in abiotic environmental conditions that constrain the onset or duration of the annual period of biological activity in seasonal environments, or through loss of species (Wolkovich and Cleland 2014). Such vacancies may be filled by phenological expansion of remaining species, or through invasion by exotic species able to position themselves phenologically into such vacancies along the temporal niche axis. Hence, species addition through invasion of absolute phenological niche space occurs through stochastic formation of niche vacancies as existing species experience increased phenological segregation in time (figure 4.1a). However, there may also be advantages to individuals of existing species in expanding into increased temporal niche space as a means of reducing intraspecific competition in time (chapter 6). In such instances, successful establishment of exotic invasive species would depend upon the outcome of interspecific interference interactions in time.

In addition to stochastic occurrence of vacancies in the absolute phenological niche, systematic vacancies may arise if early- or late-season abiotic environmental constraints on, respectively, the onset or the end of the annual period of biological activity are alleviated through, for instance, climate change (Wolkovich and Cleland 2014). An increase in the availability of temporal niche space early in the season through spring warming may promote invasion by early-season life history strategists via so-called priority effects (figure 4.1b). The successful establishment of exotic species through priority effects owing to increased availability of early-season absolute time thus depends upon outpacing phenological advances by existing species in the local assemblage. Successful phenological niche invasion through priority effects may also depend upon whether existing species exhibit phenological niche conservatism—that is, whether the phenologies of existing early-season species remain unchanged despite the increase in availability of early-season time (Post 2013). The successful invasion of eastern

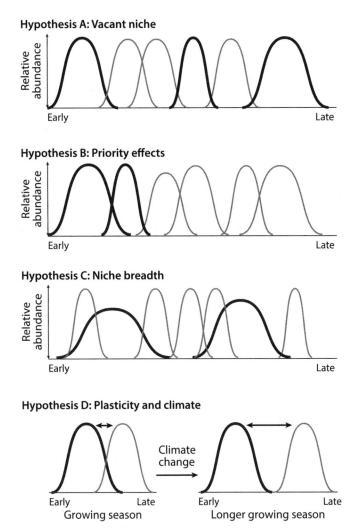

Hypothesis A: Vacant niche

Hypothesis B: Priority effects

Hypothesis C: Niche breadth

Hypothesis D: Plasticity and climate

Climate change

Early   Late   Early   Late
Growing season   Longer growing season

FIGURE 4.1. Heuristic models of phenological niche dynamics that accord with the concept of the absolute phenological niche. These were proposed as hypotheses explaining the possible ways in which exotic species (black) may invade communities of native species (gray) by exploiting vacant phenological niches (a), by exhibiting priority effects (b), through advantages of greater niche breadth (c), and by exploiting increases in the availability of early growing season time in response to climatic warming (d). Modified from Wolkovich and Cleland (2011).

North American forest understory by the invasive garlic mustard (*Alliaria petiolata*) appears to be consistent with the predictions of the hypothesis of phenological priority effects because this species tends to initiate spring growth and achieve maximum cover earlier than native species in the same local assemblages (Engelhardt and Anderson 2011).

The availability of absolute time may also increase late in the season through, for instance, prolonged autumn warming and delayed onset of freezing. Under such conditions, latency effects may apply. In this case, species characterized by late-season life history strategies, so-called autumnal species (Molau 1997), may successfully invade vacant end-of-season temporal niche space (Wolkovich and Cleland 2014). Some existing autumnal species may, however, delay or prolong their phenologies under such conditions (Sherry et al. 2007). Hence, as with priority effects, the success of phenological niche invasion through latency effects may also depend upon whether existing late-season species within the local assemblage exhibit phenological niche conservatism.

The niche breadth hypothesis posits that the establishment of exotic species may be facilitated by greater phenological niche breadth in such species than in existing species (Wolkovich and Cleland 2011, 2014). In this context, phenological niche breadth relates to duration of phenophases such as, in plants, the length of the annual period of flowering or root expansion. This hypothesis appears to presume either the prior existence or formation of vacancies in the absolute phenological niche, depicted in figure 4.1c as gaps along the seasonal time axis. Exotic species may establish in these gaps, if the availability of absolute time is amenable to their seasonal life history strategies, and then persist or possibly even displace existing species by virtue of competitive superiority deriving from greater phenological niche breadth (Wolkovich and Cleland 2011, 2014). The results of a six-year experiment conducted at the Cedar Creek Natural History Area involving introduction of the bunchgrass *Schizachyrium scoparium* appear consistent with this prediction (Fargione and Tilman 2005). Based on the timing of peak rates of soil nitrate reduction on experimental plots, *S. scoparium* filled the midseason phenological niche within the plant community, and displaced other midseason species (Fargione and Tilman 2005). In contrast, species with early-season phenologies fared well in coexistence with *S. scoparium*, presumably owing in part to phenological niche differentiation, although differential rooting depth was also implicated (Fargione and Tilman 2005).

The niche breadth hypothesis should apply equally as well to the implications of climate change for competition, coexistence, and exclusion among native species in existing local species assemblages. In such cases, we might expect species with the broadest phenological niches to fare best under climate change scenarios that increase the length of the growing season. In communities of coexisting warm-dry tolerant and cool-moist tolerant eucalyptus species, for instance, species that exhibit the widest phenological niches, such as *Eucalyptus microcarpa*, are projected to outcompete co-occurring species with narrower phenological niches (Rawal et al. 2014).

Last, the phenological plasticity hypothesis derives predictions based upon the comparative capacities of exotic and native species to respond phenologically to interannual variation and trends in climatic conditions (Wolkovich and Cleland 2011). Presumably, the capacity in some species for rapid adjustment of the timing of expression of life history traits related to growth and development, reproduction, and survival confers a selective advantage over species lacking such plasticity. This advantage should apply equally to the invasion of or expansion into phenological niche gaps and to adjustment to advances in the onset of the season of biological activity (figure 4.1d). Hence, phenological plasticity should be predictive of the capacity of species to respond adaptively to climate change. Results deriving from analyses of long-term data on the timing of emergence by plants in the local community at the study site near Kangerlussuaq, Greenland (Post 2013), may be consistent with the predictions of this hypothesis. For instance, species-specific trends in emergence timing over the period of observation (2002–2013) scale inversely with species-specific interannual variability in emergence dates (figure 4.2a). Hence, the more variable a species' emergence timing is among years, the more rapidly its emergence timing has advanced at the site. As well, species-specific trends in emergence timing scale negatively with the species-specific magnitude of weather effects on emergence phenology (figure 4.2b). Thus, species exhibiting the greatest interannual variability in emergence phenology are also those that have undergone the most rapid advances in emergence phenology over the past 12 years, presumably because they are the most phenologically responsive to variation in abiotic conditions constraining emergence timing.

THE RELATIVE PHENOLOGICAL NICHE

In this section, we will turn our attention to the timing of phenological activity by the individual organism in relation to two interacting factors: the seasonal timing of favorability of abiotic conditions, and the seasonal timing of favorability of biotic conditions. An important distinction in this development of the relative phenological niche concept, in contrast to that of the absolute phenological niche concept, thus concerns the nature of timing in each of them. Here, timing is not absolute, as in "early" or "late" along the axis of cosmological time. Rather, it is relative to the timing of other components of the system that may, themselves, vary unpredictably or predictably. The former includes stochastic variation in the timing of abiotic and biotic conditions, while the latter includes trends in the timing of abiotic and biotic conditions influencing the success of individual organisms as measured ultimately by their contributions to future generations.

Accordingly, a key distinction between the absolute phenological niche and the relative phenological niche arises from this contrast. The absolute phenological

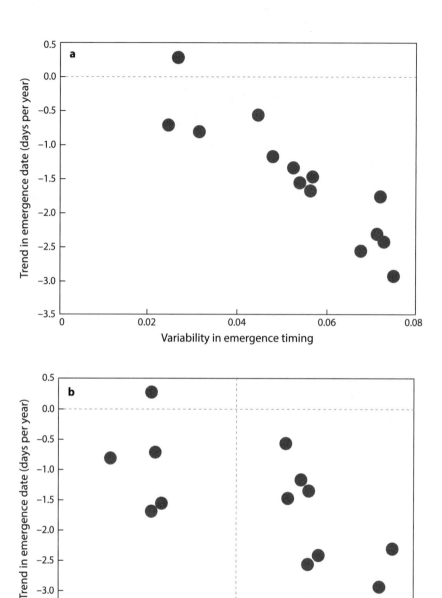

FIGURE 4.2. Associations among variability in emergence timing, trends in emergence phenology, and abiotic (weather) influences on emergence phenology in the plant community at the study site near Kangerlussuaq, Greenland, between 2002 and 2013. In both panels, each point represents metadata for an individual species. (a) Species characterized by more variable emergence timing among years (the coefficient of variation, CV, of emergence dates for the period) exhibit greater trends toward earlier emergence timing ($r = -0.91$, $P < 0.001$); and (b) these trends tend to scale with stronger influences of abiotic conditions on emergence timing ($r = -0.66$, $P = 0.02$). From Post et al. (2016).

niche may be considered to be akin to the Grinnellian niche (Grinnell 1917; Soberon 2007; Soberon and Nakamura 2009). Much as Grinnell described the distribution and occurrence of the California thrasher (*Toxostoma redivivum*) in relation to abiotic and other environmental variables (Grinnell 1917), the absolute phenological niche describes the life history characteristics of species according to their distribution and occurrence in relation to cosmological time. Put simply, the absolute phenological niche describes the temporal habitat requirements of species. In contrast, the relative phenological niche is more analogous to the Eltonian niche concept (Elton and Miller 1954; Elton 1958; Levine and D'Antonio 1999; Soberon 2007). Elton's concept of the niche focused on the functional role of the species in relation to other members of the ecological community with which it interacts. Similarly, the concept of the relative phenological niche emphasizes the species' use of time in relation to the use of time by other components of the biological system in which it is embedded.

The relative phenological niches of species may be expressed as encompassing two relative temporal dimensions: the seasonal timing of abiotic conditions favorable to growth, development, and reproductive activity of the individual, and the seasonal timing of biotic conditions favorable to them (Iwasa and Levin 1995; Post 2013; Park 2017). We may represent both the probability of encountering favorable abiotic conditions by the individual and the probability of its encountering favorable biotic conditions as Gaussian functions in relation to time (Post 2013). We may assume there is a peak probability of encountering favorable abiotic conditions and a peak probability of encountering favorable biotic conditions. We may furthermore assume that the daily probabilities about these peaks are approximately normally distributed through time, on a seasonal scale, about these peaks. Hence, these may be represented by seasonal curves with, in the simplest scenario, a single peak for the timing of favorable abiotic conditions and a single peak for the timing of favorable biotic conditions (figure 4.3).

In this setting, for any individual of a given species, prevailing abiotic conditions early in the annual season of activity are highly unfavorable, but soon reach peak favorability, followed by a decline toward unfavorable conditions later in the season (figure 4.3a). Such conditions may be represented by, for example, temperature in high-latitude systems, precipitation or solar irradiance in low-latitude systems, or some combination of these. Hence, the shape of the curve representing the function describing the relationship between time and the probability of experiencing favorable abiotic conditions will likely vary between low-latitude and high-latitude environments. For instance, the timing of the seasonal onset and the timing of the seasonal decline in the probability of encountering favorable abiotic conditions are likely to occur later and earlier, respectively, in high-latitude environments than in low-latitude environments, producing a narrower or tighter

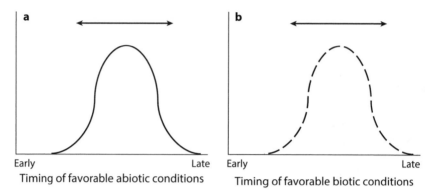

FIGURE 4.3. Development of the concept of the relative phenological niche begins with proposed seasonal variation in the timing of (a) favorable abiotic and (b) biotic conditions. During the early part of the annual season of activity (left end of each x-axis), both abiotic and biotic conditions are presumed to be unsuitable for growth, development, or reproduction. Conditions peak in favorability, and then decline again late in the season. The arrows in each panel indicate that the annual timing of peak favorability for both abiotic and biotic conditions may fluctuate from year to year, or undergo systematic shifts owing to trends in conditions.

distribution about the seasonal peak. This is, of course, a product of the Earth's tilt on its axis and its orbit around the Sun, as discussed in chapter 2.

The factors determining the timing of favorable biotic conditions will, for our purposes here, comprise the two most basic forms of organism-organism interactions: interference interactions and exploitation interactions. These will be further partitioned into competition, whether intra- or interspecific in nature; resource-consumer interactions; and consumer-resource interactions. Let us consider each of these, and their roles in determining the timing of favorable biotic conditions, in turn.

As with the timing of favorable abiotic conditions, competition for limiting resources can be expected to be minimal early in the season of biological activity, but increase rapidly to a seasonal maximum as both conspecific and heterospecific competitors become active, after which it can be expected to decline again as conditions become unfavorable for biological activity. Accordingly, the onset, peak occurrence, and termination of resource competition follow a seasonal pattern similar to that of the timing of favorable abiotic conditions (figure 4.3b). Here again, we may draw contrasts between low- and high-latitude systems according to which the shape of the competition curve may vary. One of the most obvious contrasts between low- and high-latitude systems that relates to competition concerns diversity, which declines with increasing latitude (Gaston 2000; Pimm and Brown 2004; Orme et al. 2005; Gotelli et al. 2009). In high-latitude environments, encounters among organisms are more likely to involve conspecifics, whereas in low-latitude systems they are more likely to involve heterospecifics (Gotelli and McCabe 2002). Hence, in high-latitude systems, the immediate competitive environment is likely to be dominated more by conspecific than by heterospecific individuals, and these, in comparison to heterospecifics, are more likely to exhibit nearly identical phenologies. Consequently, the shape of the curve representing the function describing the relationship between time and the probability of experiencing adverse resource competition will likely be narrower in high-latitude than in low-latitude systems. We will return to this point later in a discussion of the implications of this for relative phenological niche breadth. Last, both curves representing the seasonal onset, progression, and decline in favorability of abiotic and biotic conditions may shift, expand, or contract along the x-axis in figure 4.3 in response to fluctuations or trends in the timing and duration of the onset, peak, and decline in favorable conditions.

The curve representing the seasonal timing of the onset, progression, peak occurrence, and subsequent decline in favorability of abiotic conditions and that describing analogous dynamics relating to competition can reasonably be assumed to overlap at least somewhat. These curves will overlap considerably if abiotic conditions drive the seasonal timing of activity by the species involved, with the competition curve lagging slightly behind the curve representing favorable abiotic conditions (figure 4.4a). As depicted in this manner, then, it is the difference between the period of onset, progression, and peak timing of favorability of abiotic conditions and the period of onset, progression, and peak timing of competitor abundance that determines the competition component of the relative phenological niche (figure 4.4a, shaded region). This conceptualization may be improved if the period preceding the peak of resource competition is considered to be the period of favorable biotic conditions. Accordingly, the shaded region in figure 4.4 encompasses the period when conditions in both the abiotic environment and the biotic environment are favorable.

Next, let us examine the seasonal timing of the onset, progression, peak, and decline in favorability of abiotic conditions in relation to that describing the seasonal dynamics of resource species abundance. Just as with competitors, we can expect the phenological dynamics of resource species upon which the focal organism depends to display approximately Gaussian dynamics (figure 4.4b). Although the focus in this example is on the phenological dynamics of resource species, the same or similar dynamics apply to abiotic resources such as nutrients. Here, it is the period of *overlap* between favorable abiotic conditions and resource phenology that contributes to the relative phenological niche of the focal organism (figure 4.4b, shaded region) as the seasonal period of net positive conditions.

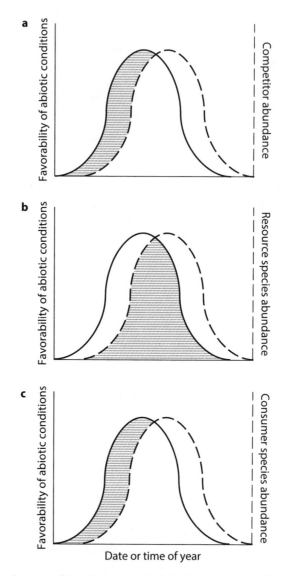

FIGURE 4.4. Development of the relative phenological niche concept as a function of a simple scenario including only two temporal niche axes: seasonal variation in the timing of abiotic conditions favorable to reproductive activity (solid line), and seasonal variation in the favorability of biotic conditions (dashed line) owing to seasonal variation in abundance of (a) conspecific and aspecific competitors, (b) resource species, and (c) consumer species. The relative phenological niche of a species (shaded area in each panel) may be thus be depicted as the temporal space during which all types of conditions are favorable and their favorability overlaps.

Last, the seasonal dynamics of consumer phenology may be compared to the seasonal timing of the onset, progression, peak, and decline in favorability of abiotic conditions (figure 4.4c). In this instance, we are concerned with the seasonal timing of appearance in the local assemblage, rise to peak abundance, and gradual retreat, of individuals of species that act as consumers of the focal organism. Hence, here again, the period of net favorability of conditions for the focal organism is defined by the *difference* between the period preceding the decline in favorability of abiotic conditions and the period preceding the peak occurrence of consumer species (figure 4.4c, shaded region).

Before proceeding toward an integration of these three components into the relative phenological niche, let us return briefly to consideration of patterns of variation in them across broad environmental and ecological gradients such as latitude. The magnitude of the difference between the two curves in figures 4.4a and 4.4c describing abiotic favorability in relation to competitors and consumers relates to the temporal offset, or lag, between the timing of onset and the rate of progression of favorable abiotic conditions and those of competitors and consumers from the perspective of the individual organism. The timing of onset, rate of progression, peak, and decline in the presence of competitors and consumers relative to that of favorable abiotic conditions can be expected to vary latitudinally. Likewise, the timing and dynamics of resource species phenology (figure 4.4b) in relation to the timing of favorable abiotic conditions, and hence the degree of overlap between them, should vary with latitude.

In general, we should expect greater overlap between the curves representing the seasonal timing of favorable abiotic conditions and those representing the phenology of competitors, resources, and consumers, in high-latitude than in low-latitude environments (figure 4.5). Accordingly, we can expect relatively narrower temporal displacement between the abiotic conditions curve and the competitor and consumer phenology curves, but relatively greater overlap between the abiotic conditions curve and resource phenology curve, in high-latitude than in low-latitude environments (figure 4.5, shaded regions). This is because abiotic environmental conditions, and the consequent availability of time for biological activity, are more highly constrained in high-latitude than in low-latitude systems (*sensu* Park 2017).

Because the relative phenological niche comprises the overlap between the timing of net favorable biotic conditions and the timing of favorable abiotic conditions (figure 4.6), it is tempting on this basis to posit that there should exist a latitudinal gradient in relative phenological niche breadth. However, the extent to which this can be expected is unclear. While absolute phenological niches may, in plants for instance, be broader in low-latitude than in high-latitude systems, the relative phenological niche is defined by timing relative to other organisms.

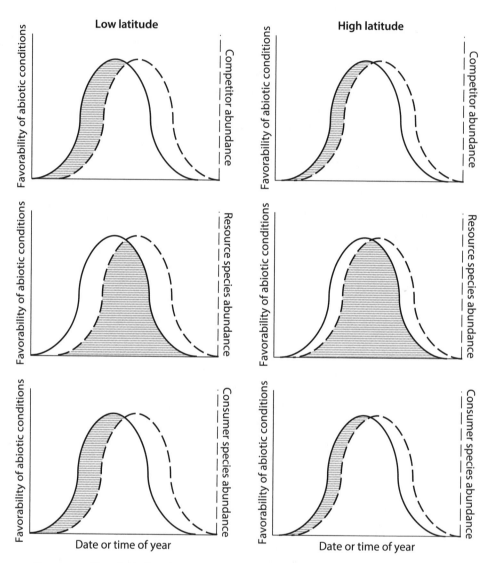

FIGURE 4.5. Hypothesized variation in the timing and extent of the seasonal period of over-lap between favorable abiotic and biotic conditions that define the relative phenological niche between low-latitude (left) and high-latitude (right) environments. Solid and dashed line curves are the same as in figure 4.4.

Hence, a shorter period of biological activity in high-latitude systems does not in and of itself constitute a narrower relative phenological niche if that period coincides with proportionally greater temporal overlap with resource species and lower temporal overlap with competitor and consumer species. Nonetheless, the differences between low- and high-latitude systems conjectured in figure 4.5 do

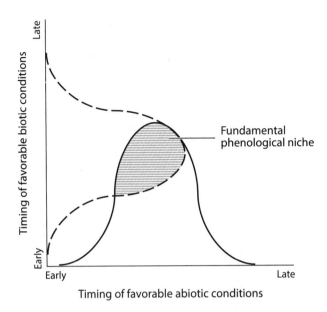

FIGURE 4.6. Rotational juxtaposition of the curves depicting seasonal variation in the timing of favorable abiotic and biotic conditions to derive the fundamental relative phenological niche space.

suggest at least one potential key difference between organisms in the two types of environment. Of the three components the relative phenological niche comprises, the greatest contribution to its breadth in high-latitude environments derives from resource phenology, while in low-latitude environments relative phenological niche breadth would appear to be driven mainly by temporal disassociation from heterospecific competitors and consumers.

Let us return now to figure 4.3, which displays curves representing the timing of favorable abiotic conditions in panel (a) and favorable biotic conditions in panel (b). We can assume for simplicity that panel (a) represents the net favorability of abiotic conditions through time, while recognizing that this allows for the possibility that multiple abiotic constraints may act simultaneously upon the individual's phenology. Likewise, we can assume that panel (b) represents the net favorability of biotic conditions through time, after accounting for potentially simultaneous constraints imposed by timing of activity by competitors, resources, and consumers as in figure 4.5. Hence, the timing of favorable abiotic conditions and the timing of favorable biotic conditions represent phenological niche axes that together compose dimensions of an organism's fundamental relative phenological niche. To represent the fundamental relative phenological niche, we thus need to achieve a superimposition of the two curves in figure 4.3 to represent their region of overlap in two dimensional niche space in which both axes represent time. To do this, we rotate panel (b) in figure 4.3 so that its time axis is now vertical, and superimpose it on panel (a) in figure 4.3. In doing so, however, we must also flip panel (b) so that the "early" ends of both time axes now meet (figure 4.6).

The resulting region of overlap between the two curves, the intersection between the timing of net favorable biotic conditions and the timing of net favorable abiotic conditions, consequently represents the fundamental relative phenological niche (figure 4.6, shaded region). This represents the time of year (or season) when growth, maintenance, and positive contributions to future generations are possible. The extent to which the fundamental relative phenological niche space is filled by a species will depend, however, on prevailing abiotic environmental conditions and the presence or absence of potential competitors (Ramos et al. 2014) as well as consumers. The relative phenological niche of a species in this idealized scenario thus results from selection at the individual level for optimal seasonal timing and duration of activity related to growth, reproduction, or survival in response to the timing of abiotic and biotic factors (Iwasa and Levin 1995). Such timing might be early or late, and the resulting phenophases might be short or prolonged, in both absolute terms and relative to timing by interacting organisms. This perspective emphasizes that phenological advance, stasis, and delay can all be viewed as components of an adaptive strategy to optimize the interrelated aspects of timing and duration of critical life history phases (see figure 3.3).

The phenological niche of the individual thus encompasses the period from increasing to peak favorability of abiotic conditions, and the period from low-onset leading up to but not including peak presence of competitors and consumers. Notably, the phenological niche is thus defined by both elements of time: absolute, cosmological time represented by the day of year or season, and relative, ecological time represented by the onset and progression of availability of other resources and activity of other species with which the individual interacts.

Employing this concept of the phenological niche, we may now further partition it into fundamental and realized elements (Connell 1961). Recall that the fundamental phenological niche of a species consists of the intersection in a two-dimensional temporal plane of the range along one axis of the seasonal timing of abiotic conditions favorable for reproductive activity, and the range, along the other axis, of the timing of biotic conditions under which resource competition, resource acquisition, and consumer avoidance are favorable (figure 4.6, shaded region). The temporal space so delineated thus represents the *potential* range of conditions under which contributions to future generations are possible.

The intersection in time between net favorable biotic conditions and abiotic conditions thus sets the bounds for the fundamental relative phenological niche, shown in a further idealized manner as the black oval in figure 4.7. Here, we have simply removed the nonoverlapping sections of figure 4.6 to emphasize the region

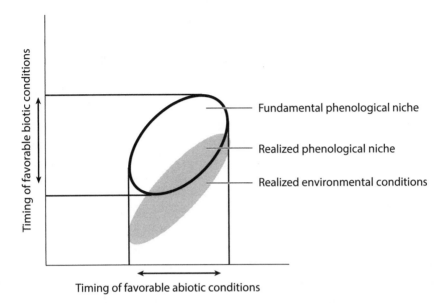

FIGURE 4.7. Relationships among the fundamental relative phenological niche of a species (heavy black line oval), the realized environmental conditions encountered by that species at the time and place it occurs (shaded oval), and the realized relative phenological niche of that species as defined by the overlap between its fundamental phenological niche and realized environmental conditions. The niche axes composing the relative phenological niche include the seasonal timing of abiotic conditions favorable for activity, development, and reproduction (*x*-axis) and the seasonal timing of favorable biotic conditions (*y*-axis) defined as shown in figure 4.4.

of overlap between favorable timing of biotic and abiotic conditions. Occupation of the entirety of this fundamental relative phenological niche by the individual organism, or by the species on average, is expected, however, to be limited by the abiotic and biotic conditions that actually prevail at any given site occupied by the species at any given time (*sensu* Williams and Jackson 2007). The timing of such conditions may reasonably be assumed to fluctuate interannually or undergo periodic trends (figure 4.3). The intersection in time between the actualized biotic and abiotic conditions favorable to growth, survival, and reproduction by the individual thus represent the realized environmental conditions at that time and place (figure 4.7 shaded oval).

This conceptualization of the realized environmental conditions differs from that employed in other demonstrations of the role of the realized environment in constraining the fundamental niche. Here, we consider the realized range of the *timing* of conditions, rather than the range in conditions themselves. Thus, the realized phenological niche of a species represents the subset of its fundamental relative phenological niche that intersects with realized environmental conditions at a location (figure 4.7). Importantly, the dimensions of the realized environment may be altered along either axis, such as through the loss or gain of species in the local assemblage or through climate change alteration of the timing of onset or duration of favorable abiotic conditions. Changes in one of these alone will primarily alter the dimensions or position of the realized environmental conditions along one axis, while changes in both may additionally shift the position of the realized environment entirely in relation to that of the fundamental phenological niche.

Later, we will explore the consequences for species interactions in phenological niche space of such shifts in realized environmental conditions resulting from climate change. The focus for the remainder of this chapter will be on mutualisms involving pollinators and plants, but should apply to any mutualism in which interactions are constrained by the phenology of each of the interacting species. With some modification, the perspectives developed later could also be extended to other types of species interactions constrained by coupled phenological dynamics, such as parasitoid-host interactions.

In each of the ensuing scenarios, it will be assumed that the relative phenological niche is highly conserved for species in both horizontal and vertical associations. In other words, it will be assumed that the boundaries of abiotic and biotic axes along which the phenological niche is represented have been fixed for any given species by selection for optimal seasonal timing of activity related to growth, reproduction, and survival. This is not to say that shifts in the timing of seasonal activity will not occur in species associations or interactions among them in response to changes in abiotic conditions or in response to changes in the biotic environment resulting from species turnover. Rather, such shifts are presumed to occur primarily in response to changes in the realized environmental conditions that represent existing abiotic and biotic conditions encountered by a species at any given site occupied by it (Williams and Jackson 2007; see figure 4.7). We will next explore such dynamics in the context of mutualisms, with a focus on the phenology of insect pollinators and plants. In the hypothetical scenarios that follow, changes in realized environmental conditions will arise, first, through effects of climate change on the timing of favorable abiotic conditions, and, subsequently and with increasing complexity, through effects of changes in resource phenology on the timing of favorable biotic conditions.

MUTUALISMS AND THE RELATIVE PHENOLOGICAL NICHE

Considerable focus has been placed recently on declines in insect pollinator abundance, species richness, and pollination efficacy, with multiple contributing factors having been implicated directly or indirectly in such declines. These include loss of floral resources and diversity through agricultural and other land use changes, pesticide use, spread of pathogens, climate change, and their interactions (Brittain et al. 2010; Potts et al. 2010a; Dutka et al. 2015; Stanley et al. 2015a, 2015b; Koh et al. 2016). Quantification of the magnitude and extent of global insect pollinator declines has been problematic, but regional assessments hint at the scope and potentially dire scale of the problem.

Among domestic insect pollinators, the honeybee (*Apis mellifera*) has undergone the most dramatic regional declines. Across Europe, for instance, honeybee stocks declined by 25 percent between 1985 and 2005, while in the United States they declined by 59 percent from 1947 to 2005 (vanEngelsdorp et al. 2008; Potts et al. 2010a, 2010b). Rates of decline among wild insect pollinators have been more difficult to estimate, but the bee species group under greatest decline appears to be bumblebees (*Bombus* sp.) (Potts et al. 2010a). In this group, up to 10 of 16 species are in decline in the United Kingdom, for instance (Williams and Osborne 2009), while in the United States the *Bombus* species complex has suffered a 30 percent decline in species richness over the past 140 years (Bartomeus et al. 2013). More generally, it has been estimated that during the five-year period between 2008 and 2013, bee pollinator abundance declined by 23 percent across the United States (Koh et al. 2016). Among other wild insect pollinators, there is evidence that hoverflies (or syrphid flies) may be increasing in some areas, such as the Netherlands, where their increases coincide with declines in bee diversity (Biesmeijer et al. 2006). At one High Arctic site in Greenland, abundance of midges and flies (Chironomidae and Muscidae, respectively) has declined in association with shorter flowering seasons of their floral resources (Høye et al. 2013). Relative declines in insect pollinator richness have also been documented, including, for example, shifts in the relative abundance of 56 percent of 187 native species of bees in the United States since the late 1800s (Bartomeus et al. 2013).

While identification of generalized drivers of such declines has been elusive, and may be beyond expectation, several interesting patterns are evident that have relevance to phenological niche conservatism (Post and Avery in press). For instance, bee species in the United States that have declined in relative abundance also display narrow dietary and phenological niche breadth (Bartomeus et al. 2013). As mentioned earlier, syrphid flies have increased in parts of the Netherlands where bees have declined. However, where both have declined in the Netherlands and United Kingdom since 1980, such declines appear to be more common among syrphid fly and bee species with specialist, rather than generalist, pollination niches (Potts et al. 2010a). Likewise, in the Netherlands, declines in pollinated species of plants have been more common among bee-pollinated than among syrphid fly–pollinated species (Potts et al. 2010a). A long-term observational study conducted at the Rocky Mountain Biological Laboratory in Colorado, between 1992 and 2011, revealed that syrphid fly pollinators were generally able to maintain overlap with floral resource phenology despite more rapid responses by resource plants to the alleviation of seasonal abiotic constraints on flowering phenology (Iler et al. 2013c). In this study, the key driver of the onset of both flowering phenology and syrphid emergence was the annual timing of snow melt at the site, with syrphid flies tending to emerge subsequent to the onset of flowering. The maintenance of overlap with floral resource phenology in this instance likely relates to phenological niche generalization among syrphid flies and their consequent ability to make use of multiple species of flowering plants (Iler et al. 2013c).

The differential biodiversity consequences of climate change for phenological niche specialists and generalists will be explored in this and the following section. The tremendous array of ecosystem services rendered by insect pollinators in natural and agricultural systems is undoubtedly susceptible to alteration by declines in pollinator diversity (Kearns et al. 1998; Klein et al. 2007; Kremen et al. 2007; Gallai et al. 2009; Koh et al. 2016), as well as through disruption of these mutualisms through phenological decoupling (Petanidou et al. 2014). As in the study by Iler et al. (2013c), flowering plants may generally (Solga et al. 2014), though not universally (Bartomeus et al. 2011), advance their phenology more rapidly in response to alleviation of abiotic constraints through climatic warming than do their insect pollinators. In part, this is attributable to the requirement among insect pollinators for higher temperatures and greater accumulation of degree days before activity is triggered compared to flowering plants (Forrest and Thomson 2011). And, while specialist insect pollinators are likely to suffer adverse consequences of resource loss through such phenological mismatches (Hegland et al. 2009), specialist plants are not (Petanidou et al. 2014). This is because specialist plants are capable of pollination by multiple pollinator species, though fewer than generalist plants (Petanidou et al. 2014), and generalist pollinators may replace specialist pollinators if the phenology of the latter does not keep pace with that of their floral resource species (Solga et al. 2014).

Indeed, pollinator species that are likely to persist under climate change advancement of phenology are those characterized by the capacity for rapid phenological response and by generalist mutualisms (Burkle and Alarcon 2011). For instance, bumblebee species suffering range declines in the United Kingdom are those with narrow climatic niches (Williams et al. 2007; Williams and Osborne 2009). Likewise, in parallel with pollinator declines, loss of pollination services through phenological mismatch may increase pollen limitation and reduce reproductive success in plants (Thomann et al. 2013). Hence, generalist pollinator-plant mutualisms may be less vulnerable to phenological disruption owing to climate change (Gilman et al. 2012) or even to alterations in the timing of favorable biotic conditions. Now let us examine, in turn, the implications of phenological niche conservatism in specialist and generalist pollinator-plant associations by exploring scenarios relating to shifts in the timing favorable abiotic and favorable biotic conditions.

BIODIVERSITY IMPLICATIONS OF PHENOLOGICAL NICHE CONSERVATISM IN SPECIALIST ASSOCIATIONS

In specialist pollinator-plant associations, it may be reasonably assumed that the phenological niches of both plant and pollinator overlap fairly closely with each other and with environmental conditions (Shapiro 1973). In such a situation, let us first consider a specialized single pollinator–single plant system. Here, the term *specialist* is applied from the perspective of the pollinator. Hence, the plant species may persist despite changes in the scenarios that follow, provided realized environmental conditions overlap at least partially with its fundamental phenological niche. Persistence of the specialist pollinator, however, requires maintenance of overlap of its phenological niche with that of the plant species and with realized environmental conditions. Loss of overlap with either one, therefore, is assumed to result in local extinction of the pollinator.

In the scenario depicted in the pair of panels in figure 4.8a, the fundamental phenological niche of the plant (dashed oval) is encompassed almost completely by realized environmental conditions (the shaded oval). The fundamental phenological niche of the specialized pollinator (black oval) is, however, offset slightly from that of the plant along both the y-axis, representing the timing of favorable biotic conditions, and along the x-axis, representing the timing of favorable abiotic conditions. The x-axis offset suggests that the timing of abiotic conditions favorable for pollinator activity is slightly delayed compared to the timing of abiotic conditions favorable for its plant mutualist. This is intended to comport with the observation common in empirical studies of earlier seasonal activity among flowering plants than among their insect pollinators (Iler et al. 2013c; Kudo and Ida 2013; Forrest 2015). The zone of overlap among the realized environmental conditions, the fundamental phenological niche of the plant, and the fundamental phenological niche of the pollinator, represents the window of successful phenological mutualism in this association.

Now let us examine the consequences for this mutualism of shifts in the availability of time along both axes. Climate change should result in a shift in realized environmental conditions along the axis representing the timing of favorable abiotic conditions in either direction. This would result in either advanced timing of favorable conditions (leftward shift), delayed timing of favorable conditions (rightward shift), or prolonged timing of favorable conditions (expansion).

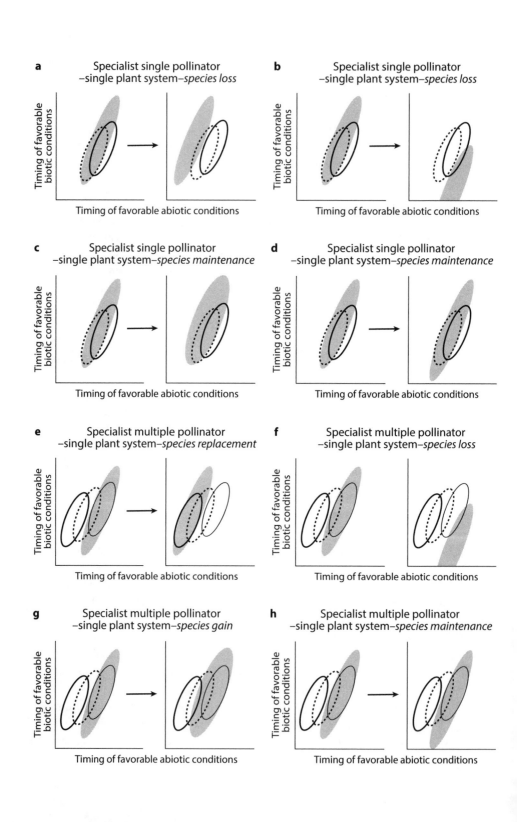

Changes in the seasonal timing of activity of competitors or consumers should, on the other hand, result in shifts in realized environmental conditions along the y-axis. This would result in either advanced timing of favorable conditions (downward shift), delayed timing of favorable conditions (upward shift), or prolonged timing of favorable conditions (expansion). For simplicity, we will examine shifts toward earlier timing of favorable conditions and prolonged timing of favorable conditions. We could, however, just as well examine the consequences of shifts toward delayed timing of conditions or reduced duration of them using the same approach. Note, also, that prolongation of the timing of favorable conditions in each of the scenarios depicted in the following results from an advance toward earlier timing of favorable conditions coupled with maintenance of baseline end-of-season timing. The arrows in each panel of figure 4.8 represent the shift in timing of conditions from baseline to future scenarios.

In the first scenario (see figure 4.8a), climate change results in an advance in the timing of favorable abiotic conditions, through, for instance, warming or earlier snow melt. As a consequence, there is a loss of the specialist pollinator in this system as the shift in realized environmental conditions moves conditions beyond the x-axis boundary of the pollinator's fundamental niche. The plant in this case retains a narrower realized phenological niche and may persist, unless, as some documented examples of pollinator declines have demonstrated, it becomes pollen limited (Thomann et al. 2013). In the next scenario, the timing of favorable biotic conditions has advanced, and as a consequence the realized environmental conditions have now moved beyond the boundary of the fundamental phenological niche of the plant species in this mutualism (figure 4.8b). Hence, we can expect the plant to become locally extinct under these conditions. Likewise, despite the fact that the pollinator maintains a slight degree of overlap between its fundamental phenological niche and realized environmental conditions, it is expected to undergo local extinction as well. This will result, as explained earlier, from the loss of its specialist floral resource.

FIGURE 4.8. Implications of climate change for specialist pollinator-plant mutualisms as a function of changes in realized environmental conditions (shaded oval) relative to the fundamental phenological niches of interacting plant (dashed line oval) and pollinator (solid line oval) species. In each pair of figures, the left panel represents current conditions, the arrow represents climate change, and the right panel represents future conditions resulting from climate change. Pairs (a) through (d) represent mutualisms involving a single pollinator, while pairs (e) through (h) represent mutualisms with multiple potential pollinators. It is assumed in all cases that niche dimensions are conserved. Pairs (a), (c), (e), and (g) depict changes in the timing of favorable abiotic conditions, while pairs (b), (d), (f), and (h) represent changes in the timing of favorable biotic conditions.

In the second scenario, under a prolongation of the timing of favorable abiotic conditions, the specialist pollinator-plant mutualism is preserved because realized environmental conditions still encompass the entire temporal window of the phenological mutualism (figure 4.8c). Hence, both species may be expected to persist. Moreover, such an expansion of realized environmental conditions along the abiotic timing axis may promote invasion of the mutualism by a second pollinator species. This might be expected to result if an increase in the range of timing of favorable abiotic conditions at the site has resulted in the opening of additional phenological niche space. As well, a prolongation of the timing of favorable biotic conditions similarly maintains the existing mutualism because realized environmental conditions persist within the fundamental phenological niches of both plant and pollinator (figure 4.8d).

In the next set of scenarios, we will examine the consequences of the same types of advances in, and prolongation of, the timing of favorable abiotic and biotic conditions in specialist mutualisms, but in this case for multiple consumer-single resource systems. Here, we have a hypothetical system that, under baseline conditions, comprises a single plant and two pollinators. The two pollinator species in this assemblage display niche complementarity. Although their fundamental phenological niches overlap almost entirely along the axis representing the timing of favorable biotic conditions, they segregate along the axis representing the timing of favorable abiotic conditions (figure 4.8e, left panel). However, prevailing realized environmental conditions (shaded oval) in this baseline scenario are amenable only to the plant (dashed oval) and one pollinator species (thin black oval). In this case, a shift toward earlier timing of favorable abiotic conditions results in replacement of this pollinator by the other (thick black oval; figure 4.8e). Hence, persistence of the plant is possible under these novel environmental conditions, although in a new mutualism. The original pollinator is expected to suffer local extinction in this case owing to loss of overlap between its fundamental phenological niche and realized environmental conditions. Note that the driver of extinction in this instance differs from that in the scenario depicted in figure 4.8b, where extinction resulted from the loss of floral resources despite persistence of favorable environmental conditions for the pollinator.

In the next panel (figure 4.8f), we see that a shift toward earlier timing of favorable biotic conditions, with the timing of favorable abiotic conditions held constant, results in dissolution of the original mutualism and species loss. This is because realized environmental conditions have become dissociated entirely from the fundamental phenological niche of the plant species, which precipitates its extinction. Here again, as in the scenario given in figure 4.8b, the pollinator is expected to go extinct despite maintenance of some overlap between its fundamental phenological niche and realized environmental conditions because it has lost its floral resource species.

A prolongation of the timing of favorable abiotic conditions would be expected, in this system, to result in the net gain of a pollinator species (figure 4.8g). The original mutualism is preserved in this case, and warming has expanded the realized environmental conditions along the abiotic timing axis just enough to promote overlap between the plant and a second pollinator in phenological niche space. In contrast, a prolongation of the timing of favorable biotic conditions simply maintains the existing mutualism (figure 4.8h). Here, overlap between the plant and its original pollinator with realized environmental conditions is unchanged, and these conditions do not promote overlap with the phenological niche of the second pollinator species. Next, let us examine implications of such changes in the timing of favorable abiotic and biotic conditions for generalist pollinator-plant associations.

BIODIVERSITY IMPLICATIONS OF PHENOLOGICAL NICHE CONSERVATISM IN GENERALIST ASSOCIATIONS

For the purposes of the following discussion, the use of the term *generalist* in pollinator-plant associations will have dual meaning. First, it applies to the nature of the mutualistic association from the perspective of both the plant and the pollinator, and it implies that members of both groups are able to benefit from interactions with multiple species of the other group. Second, it applies to the range of abiotic and biotic conditions to which one or both of them are adapted in a temporal context. We can therefore expect that these should be broader than those to which specialists are adapted. As we examine the consequences of changes in the timing of favorable abiotic and biotic conditions, however, we will focus primarily on such consequences for pollinator persistence. In the scenarios that follow, we will observe that persistence of the pollinator may be possible despite loss of phenological overlap with mutualist plant species resulting from changes in realized environmental conditions. Pollinator persistence would, under such conditions, be promoted by a switch to alternative floral resource species, even though, for the sake of simplicity, such alternative species will not be considered later.

In the scenarios that follow, the hypothetical realized environmental conditions prevailing at a given site remain narrow, and are matched closely by the focal plant species, as in the scenarios explored earlier involving specialist pollinator-plant associations. Let us begin with consideration of a single, generalist pollinator-plant association (figure 4.9a). In contrast to the close association between the fundamental phenological niche of the plant in this association (figure 4.9a, dashed oval) and baseline realized environmental conditions (figure 4.9a, shaded oval), the fundamental phenological niche of the pollinator (figure 4.9a, black oval) encompasses a wide seasonal period of favorable abiotic conditions.

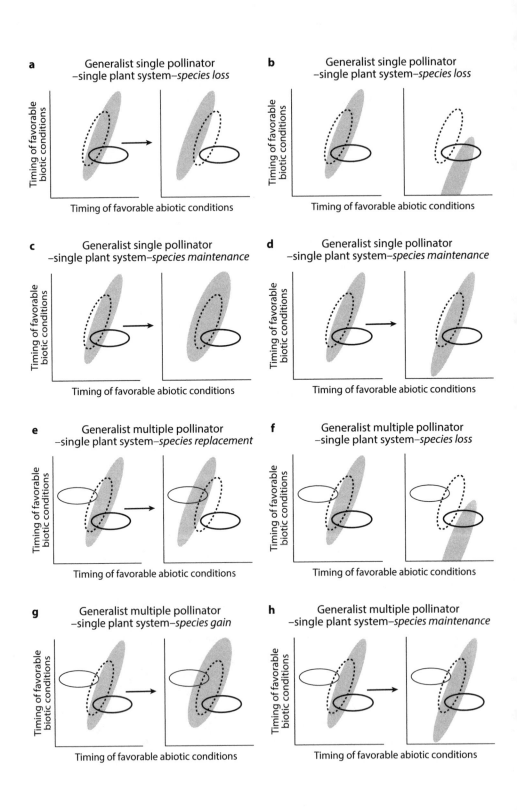

a Generalist single pollinator
–single plant system–*species loss*

Timing of favorable biotic conditions

Timing of favorable abiotic conditions

b Generalist single pollinator
–single plant system–*species loss*

Timing of favorable biotic conditions

Timing of favorable abiotic conditions

c Generalist single pollinator
–single plant system–*species maintenance*

Timing of favorable biotic conditions

Timing of favorable abiotic conditions

d Generalist single pollinator
–single plant system–*species maintenance*

Timing of favorable biotic conditions

Timing of favorable abiotic conditions

e Generalist multiple pollinator
–single plant system–*species replacement*

Timing of favorable biotic conditions

Timing of favorable abiotic conditions

f Generalist multiple pollinator
–single plant system–*species loss*

Timing of favorable biotic conditions

Timing of favorable abiotic conditions

g Generalist multiple pollinator
–single plant system–*species gain*

Timing of favorable biotic conditions

Timing of favorable abiotic conditions

h Generalist multiple pollinator
–single plant system–*species maintenance*

Timing of favorable biotic conditions

Timing of favorable abiotic conditions

However, it also comprises a comparatively narrow seasonal period of favorable biotic conditions, and hence overlaps only partially with both its plant mutualist and realized environmental conditions (figure 4.9a). The narrow range of seasonal timing of favorable biotic conditions characterizing the phenological niche of this pollinator may reflect niche complementarity along the biotic timing axis, even though no other pollinators are present at the site characterized by the baseline conditions in this scenario.

The realized environmental conditions, while overlapping the fundamental phenological niche of the plant almost completely, overlap in this instance only a narrow subset of the pollinator's fundamental phenological niche on the axis representing the timing of favorable abiotic conditions. Hence, in this case, there is initially a narrow window of phenological mutualism where all three ovals intersect (figure 4.9a). Here, just as in the example of a single specialist association (figure 4.8a), any shift owing to climate change toward earlier timing of favorable abiotic conditions for the pollinator will risk resulting in loss of the pollinator from the association (figure 4.9a, right panel).

Let us next consider the consequences for this mutualism of a shift toward earlier timing of favorable biotic conditions (figure 4.9b). In this case, realized environmental conditions at the site move quickly out of the zone of overlap between the phenological niches of the plant and the pollinator (figure 4.9b, left panel). Consequently, the mutualism is disrupted as the plant species undergoes local extinction at the site. However, the pollinator may persist. The primary difference between this outcome and that considered for the specialist, single plant–single pollinator mutualism (figure 4.8b) is that in this case realized environmental conditions now fall nearly squarely within the center of the pollinator's fundamental phenological niche (figure 4.9b, left panel).

Last, let us consider the consequences of prolonged timing of favorable abiotic conditions and of favorable biotic conditions. In both cases, prolongation occurs as a consequence of earlier timing of onset of favorable conditions with the timing

FIGURE 4.9. Implications of climate change for generalist pollinator-plant mutualisms as a function of changes in realized environmental conditions (shaded oval) relative to the fundamental phenological niches of interacting plant (dashed line oval) and pollinator (solid line oval) species. In each pair of figures, the left panel represents current conditions, the arrow represents climate change, and the right panel represents future conditions resulting from climate change. Pairs (a) through (d) represent mutualisms involving a single pollinator, while pairs (e) through (h) represent mutualisms with multiple potential pollinators. It is assumed in all cases that niche dimensions are conserved. Pairs (a), (c), (e), and (g) represent changes in the timing of favorable abiotic conditions, while pairs (b), (d), (f), and (h) represent changes in the timing of favorable biotic conditions.

of end-of-season conditions remaining consistent with baseline conditions. As figures 4.9c and 4.9d indicate, an expansion of realized environmental conditions along the axis representing favorable timing of abiotic conditions, and along the axis representing favorable timing of biotic conditions, respectively, will leave the existing generalist pollinator-plant association intact.

In associations involving multiple generalist pollinators, niche complementarity may be expected to manifest as minimal overlap among co-existing species of pollinators along one or both axes of their respective fundamental phenological niches (Heithaus 1979; Shapiro et al. 2003; Bluethgen and Klein 2011; Carvalho et al. 2014). In the following set of scenarios, we will consider the consequences of phenological niche conservatism under changing temporal conditions in multispecies generalist associations. Here, two species of generalist pollinators overlap slightly along the axis representing the timing of favorable abiotic conditions, and thus may be assumed to display sequential emergence phenology where they co-occur. However, under initial conditions, the realized environment, as in the examples in figure 4.9a–d earlier, is amenable initially only to the plant and one of these generalist pollinators (figure 4.9e, left panel). As climatic warming unfolds, and the timing of realized, favorable abiotic conditions advances, this original pollinator species is lost from the association and replaced at the site by the other species of pollinator, whose fundamental phenological niche is now favored partially by the new realized environmental conditions at the site (figure 4.9e, right panel). Obviously, such species replacement is possible only if the second pollinator is able to colonize the site once abiotic conditions become favorable for it there, such as through poleward or elevational redistribution (Parmesan 2006). A shift toward earlier timing of favorable biotic conditions results in dissolution of the mutualism and local extinction of the plant species owing to the fact that realized environmental conditions at the site now lie entirely outside of its niche space (figure 4.9f), as in the single-pollinator scenario depicted in figure 4.9b. Here again, realized environmental conditions appear amenable to the persistence of the original pollinator species.

A lengthening of the seasonal period of favorable abiotic conditions in the absence of a shift in mean conditions results in addition of the second pollinator species to the association without loss of the original pollinator; hence, there is a net species gain under climatic warming in this scenario (figure 4.9g). Such an addition of another pollinator species to the mutualism is only likely, however, if, as earlier, the species is able to colonize the site, and if the two pollinator species segregate along another axis of the phenological niche such as that representing the timing of favorable biotic conditions. Last, a lengthening of the timing of favorable biotic conditions appears to result only in maintenance of the single plant–single pollinator mutualism present in baseline conditions (figure 4.9h). Such a lengthening does not alter phenological overlap in this original association. Nor does it promote addition to this association of the second pollinator species.

What, if any, general conclusions about the consequences of phenological niche conservatism derive from these exercises? First, in both specialist and generalist associations, advances in the timing of favorable abiotic conditions and in the timing of favorable biotic conditions both resulted in dissolution of the mutualism, but for different reasons. A shift in realized environmental conditions owing to earlier onset of favorable abiotic conditions precipitated local extinction of the pollinator but favored persistence of the plant. In contrast, a shift in realized environmental conditions owing to earlier seasonal onset of favorable biotic conditions dissolved the mutualism through local extinction of the plant, while favoring persistence of the pollinator. Second, in both specialist and generalist multiple pollinator–single plant associations, shifts in realized environmental conditions owing to advancing timing of favorable abiotic conditions led to pollinator species replacement, and hence formation of novel mutualisms. Shifts in realized environmental conditions owing to advancing timing of favorable biotic conditions, in contrast, led only to dissolution of existing mutualisms through plant species loss. These insights should lead us to wonder, more generally, what the role of time is in structuring phenological communities. This will be the focus of the next chapter.

CHAPTER FIVE

The Phenological Community

Ecology of Climate Change introduced the notion of the phenological community, and drew distinctions between it and the traditional community concept concerned with species composition of the local assemblage (Post 2013). Here, that notion will be revisited briefly before an examination of how this concept relates to the utilization of time by species that co-occur in the local assemblage. Then we will examine the consequences for phenological community dynamics of differential use of time by co-occurring species. A main point of emphasis in this chapter is the dynamic nature of the community in a phenological context. The allocation of time by the individual organism to phenophases within its annual cycle of growth, maintenance, and reproduction determines patterns of interactions in time among species co-occurring in the local assemblage. In the context of phenology, the local community is characterized by a capacity for pronounced variability on both short term temporal scales, over days, and on longer-term temporal scales, from year to year.

THE COMPOSITIONAL VERSUS PHENOLOGICAL COMMUNITY

The traditional concept of community composition refers to the proportions of the total community abundance or species richness contributed by the individual species it comprises (Begon et al. 2006). In this framework, dominant species are those that have a proportionately greater abundance or occurrence, and subordinate species are those with a proportionately low abundance or occurrence. In some instances, these designations also coincide with competitive dominance and inferiority. Changes in the abundance of species through time, or in the occurrence of species through time, hence alter community composition. Generally, such changes are realized over relatively longer time scales through succession or gradual shifts in species' relative abundances. But they may also occur over the short term through species removal by local extinction or through species addition by colonization (Ricklefs 1987, 2011). Nonetheless, when we speak of the composition of a community or of the local species assemblage, we generally refer to,

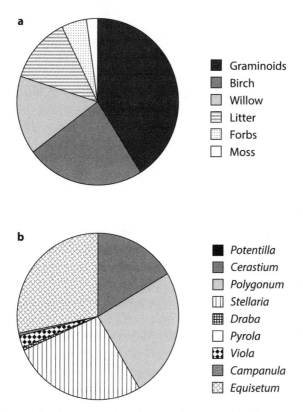

FIGURE 5.1. The traditional representation of community composition in ecology. In panel (a), the plant community at the Kangerlussuaq study site is depicted according to its proportional composition by functional groups, including graminoids, forbs, and bryophytes; as well as according to the two dominant species of dwarf shrubs, dwarf birch and gray willow; and litter. In panel (b), the composition of the forb component of the community is further broken down and depicted according to the nine genera present on the study plots the site comprises. These data derive from a single sampling period during midsummer of a single growing season in the year 2009. Hence, the picture of community composition developed on the basis of such data is highly location and time specific (Post 2013).

and present a quantification of, the relative abundances of co-occurring species in a time-static snapshot (figure 5.1). The phenological community, by contrast, refers to the proportional representation in time by co-occurring species in the local assemblage during each phenophase. Hence, the phenological community concept does not incorporate considerations of relative abundance but instead of relative phenological activity.

As an example illustrating the concept of the phenological community, and how this concept differs from that of the time-static concept of community composition, consider an Arctic plant community. When asked to characterize the composition of the community, we might conduct abundance sampling at the peak of the growing season. Doing so would result in estimates of the abundance of each species or taxonomic group which, when summed, will allow us to estimate the proportional contributions of each of these to the community (figure 5.1). In this example, we would conclude that the community is dominated by graminoids, which comprise approximately 43 percent of the total aboveground abundance, followed by dwarf birch, which constitutes approximately 25 percent of aboveground abundance. The forb component of the community, composing only 4 percent of the total community, is dominated by *Equisetum* sp. (38 percent) and *Stellaria* sp. (36 percent).

But such estimates of community composition ignore the fact that the species and taxonomic groups represented by it are characterized by differing phenological schedules and life history cycles. This is not a unique feature of high-latitude systems, and in fact partitioning of time on an annual basis by co-occurring species is evident in many different types of systems, including tropical systems (chapter 8). Multispecies assemblages of butterflies at midlatitude sites in California, for instance, display similar dynamics in the numbers of species engaged in flight activity throughout the year (Shapiro et al. 2003). Analogous, albeit shorter term, dynamics are apparent in intraseasonal variation in timing of emergence and activity of co-occurring mosquito species in high-latitude ephemeral ponds (Park 2017) and invertebrate inquiline communities inhabiting pitcher plants at midlatitudes (Ellison et al. 2003). Hence, each seasonal or annual cycle within a community is characterized by differential and varying expression of phenophases by co-occurring species. The phenological community concept thus attempts to capture representation of species by phenophases or life history stages and variation in this through the seasonal or annual cycle.

If we view the same Arctic plant community according to representation of its taxa in one phenophase, leaf emergence, we can immediately observe the dynamic nature of the phenological community (figure 5.2). As species in the local assemblage initiate growth according to their highly individualistic life history strategies, their proportional contributions to the phenological community comprising emergence varies. If we were to observe this community on day 130, we would conclude that the total number of species engaged in emergence includes only two graminoids, *Poa* sp. and *Luzula* sp. (figure 5.2). Observing the community 15 days later, we would conclude that the community comprised four species engaged in emergence: the two previously recorded graminoid species and two forbs, *Cerastium* sp. and *Draba* sp. (figure 5.2). Each species on this date thus represents one-fourth of the phenological community in a state of emergence. Not until approximately day 153 are all 11 species represented in the phenological

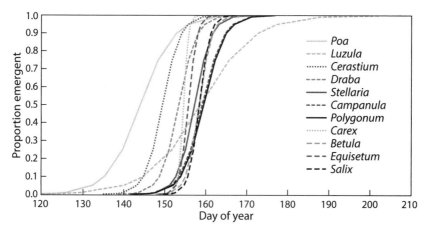

FIGURE 5.2. A representation of the concept of the phenological community. In this example, the phenological progression of emergence of each of the plant species in a community at the Kangerlussuaq study site is depicted through the course of a single growing season, 2009. The number of species engaged in a single phenophase within the phenological community changes on a near-daily basis, as the species composing the community initiate aboveground growth according to their individual life history strategies. Our estimation of the composition of the community by conventional means (for example, proportional composition as depicted in figure 5.1 for the same community in the same year) would vary depending on the exact date of sampling in this single season. For instance, sampling early in the season might produce an estimate of community composition based on the presence of only three species, while sampling late in the season would produce an estimate of community composition based on the full complement of species present in the local assemblage. From Post (2013).

community actively engaged in leaf emergence (figure 5.2). Similarly, we might record and track engagement of species in other phenophases, such as flowering, fruit production, or seed set. Doing so allows us to conceptualize the community in a dynamic fashion that captures variation among species according to their life history strategies and phenological progression. Application of the phenological community perspective also allows us to conceptualize the potential for variation among species in their allocation of time to various phenophases and the consequences of this for species interactions.

Such phenological dynamics are apparent in the sequential emergence of plant species in the local community near Kangerlussuaq, Greenland (see figure 5.2). Depending upon when during the season the local community is observed, both the species present and their relative abundances vary (figure 5.2). It can be argued, of course, that in an example such as this all of the species actually do co-occur within the local assemblage during the earliest dates of the growing season because they possess belowground biomass and interact belowground before initiating aboveground production (Harris 1977). Hence, the species in this local assemblage are potentially engaged in interference or facilitative interactions with one other early in the season through belowground dynamics even though they have not undergone emergence and production of aboveground biomass (Radville et al. 2016). However, examinations of species-specific belowground phenological dynamics reveal comparably increasing complexity of the phenological community as the season of root production commences and progresses. For instance, a study of seasonal root phenology in a community comprising bluebunch wheatgrass (*Agropyron spicatum*), cheatgrass (*Bromus tectorum*), and medusahead grass (*Taeniatherum asperum*) revealed a progression in the species-specific timing of root production, from, first, cheatgrass to medusahead grass, and last, bluebunch wheatgrass (Billings et al. 1977). Such dynamics appear comparable to the progression of the aboveground phenological community at Kangerlussuaq. The phenological community thus represents the relative timing of appearance in and progression through time of co-occurring species in the local assemblage as the season of activity unfolds for each species.

Superimposed upon the dynamics of the phenological community is interannual or longer-term variability in time. As we saw in chapter 2, rates of phenological advance vary considerably among taxa, with some species displaying substantial variability from year to year in their timing of activity, and others displaying little such variability. The consequences of such variation at the level of the local species assemblage may include shifts through time, over the course of multiple years for instance, in the composition of the phenological community. Such shifts are analogous to changes in community composition, but refer to changes in the timing of activity of individual species relative to the timing of activity by other species with which they co-occur in the local assemblage, rather than to changes in relative abundance. How might we track such shifts in the phenological community? One approach involves quantifying the seasonal rank order of appearance or activity of species in the local assemblage and monitoring changes in these from year to year.

PHENOLOGICAL COMMUNITY DYNAMICS
AND SPECIES' RANK ORDER

In seasonal environments, as resident species become active in, or as migratory species begin to arrive in, the local community, there is variation among them in their timing of onset of activity or arrival. Species with vernal life history strategies tend to become active or arrive early in the season, while those with autumnal life history strategies tend to become active or arrive later in the season (Molau

1997; Molau et al. 2005). The very existence of such a gradient in life history strategies from early-onset to late-onset species is indicative of the partitioning of time within the local species assemblage. But even though species employ an array of such strategies along this gradient, there is still some degree of variation in the timing of onset of activity or of arrival among species that we may classify as having "early" or "late" life history strategies. This variation can be quantified by assigning a rank to each species in the local assemblage corresponding to the order in which it becomes active or arrives in relation to such timing by co-occurring species (Molau et al. 2005). This is the rank order of species within the phenological community. Variation through time at scales beyond a single active season among species in their respective ranks within the phenological community would be indicative of individualistic strategies for responding to changes in the availability of ecological time (chapter 6).

If we return to the example of the timing of emergence by plants in the local species assemblage at the Kangerlussuaq study site in figure 5.2, we can see that the species are arranged in the legend according to the order in which they initiated emergence at the site that year. During the 2009 growing season, the grass *Poa pratensis* was the first species to initiate emergence, and the deciduous shrub *Salix glauca* was the last species to initiate emergence, or, in this case, leaf bud opening (figure 5.2). Additionally, we could, in this example, assign the ranks of species according to the order in which they completed emergence. Doing so would reveal that the sedge, *Carex bigelowii*, despite being one of the last species to initiate emergence, was the first species to complete emergence at the study site in 2009. By contrast, the second species to initiate emergence, *Luzula* sp., was the last to complete it that year (figure 5.2).

Over the course of the multiannual observational study of species-specific plant phenological dynamics at the Kangerlussuaq study site since 2002, it has become possible to detect patterns of interannual variation in species' rank order of emergence in the local assemblage. Such patterns indicate that the phenological community is abundantly dynamic, despite the fact that the compositional community may remain fairly fixed. As an example of such dynamics, consider the plot of the annual rank order of emergence of plant species in the local assemblage between 2002 and 2013 (figure 5.3).

The plot in figure 5.3, which depicts time series within a time series, captures three aspects indicative of the dynamic nature of the phenological community. First, species are arranged along the *y*-axis in their order of emergence relative to the timing of emergence of other species within the local assemblage. Hence, this arrangement captures each species' rank order within the phenological community. Second, the rank order series are plotted for each year for which the community has been monitored. This time series reveals the extent to which the temporal

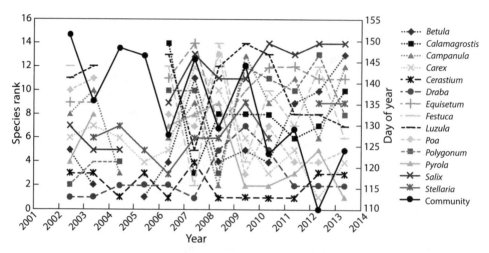

FIGURE 5.3. Variation in the rank order of emergence of plant species within the phenological community at the Kangerlussuaq study site through time based on annual sampling from 2002 through 2013. This representation of the community provides information on variation in both temporal and compositional elements of the community. In each year, individual species are ranked according to the order in which they initiate aboveground growth in the local assemblage, based on the dates on which growth is initiated by each species (not shown). This method also captures variation in the composition of the traditional community because it records variation in the presence or absence of species in the phenological record. For instance, during a major caterpillar outbreak in years 2004 and 2005, very few species were observed to initiate aboveground growth compared to nonoutbreak years. Variation in the composition of the phenological community is also evident in fluctuation and trends in the rank order of many species. From the first year of observation to the most recent year of observation, for instance, *Pyrola grandiflora* has advanced within the phenological community from the fourth to the first species to emerge. By contrast, over the same period, *Salix glauca* has regressed within the phenological community from the seventh (middle) to the fourteenth (last) species to emerge. The importance of such variation to the contribution of an individual species to the phenological community would be analogous to, in the case of *Salix glauca*, a shift in its contribution to the compositional community from midrange abundance to lowest abundance among all species in the local assemblage. Black dots plotted on the secondary (right) y-axis represent the annual date of onset of community-level emergence. Adapted from Post et al. (2016).

position of each species relative to others within the phenological community varies from year to year. Third, the rank order series for each year are plotted along the *x*-axis according to the date within each year on which community-level emergence commenced. Although the temporal resolution is too poor along the *x*-axis to detect much variation in the onset of community-level emergence, with the same data plotted on the secondary *y*-axis, a significant trend toward earlier onset of emergence at the community level is apparent (see figure 5.3).

At first glance, the most readily observable pattern in this plot is nothing more than the messy jumble of points and lines connecting them; trends are not obvious. If we consider what this jumble represents, however, the message is clear. Far from being static and predictable, the phenological community comprises species with apparently highly variable life history dynamics jostling for position in time within and among growing seasons. It is also apparent that, while the composition of the community remains relatively stable, there is variation in the representation of species among years. During the 2004 and 2005 growing seasons, a major outbreak of the caterpillar larvae of a noctuid moth, *Eurois occulta*, reduced aboveground biomass of all species of plants present at the study site (Pedersen and Post 2008; Post and Pedersen 2008), inhibiting the detection of emergence of some species entirely (figure 5.3). A second, but far less severe, outbreak occurred again in 2010 and 2011 (Avery and Post 2013) without comparable reductions in aboveground biomass or any apparent effects on our estimates of species-specific emergence dates. Aside from such compositional dynamics related to exploitation by herbivores, some species, such as the flowering forbs *Stellaria longipes* and *Polygonum viviparum*, tend to wink in and out of the phenological community (figure 5.3).

Among the 14 species monitored at the site, 5 displayed significant trends in their rank order of emergence. Of these, only one, the grass *Poa pratensis*, underwent a significant negative trend within the phenological community toward earlier rank in the timing of emergence relative to the other species in the community. By contrast, the 4 other species displaying significant trends, including the two dwarf shrubs, *Betula nana* and *Salix glauca*, and two flowering forbs, *Stellaria longipes* and *Polygonum viviparum*, underwent significant positive trends toward progressively later emergence ranks within the community. The remaining 9 species maintained relatively stable ranks within the phenological community, despite considerable interannual variation (figure 5.3).

The occurrence of positive trends toward progressively later ranks in the order of emergence does not, however, imply that the timing of emergence in such species has undergone trends toward later absolute timing of emergence. In fact, 13 of the 14 species in the local assemblage all underwent significant trends toward earlier absolute timing of emergence, and only *Salix glauca* displayed no significant trend in its absolute timing of emergence (Post et al. 2016). Furthermore, there appears to be a life history component to trends, or lack of them, in species' rank order of emergence. In other words, whether an individual species has undergone a trend toward progressively earlier or later rank in its annual timing of emergence relative to other species with which it co-occurs does not appear entirely random.

To elaborate, when estimates of the trends in individual species' rank order of emergence across the period of observation are plotted against species' mean

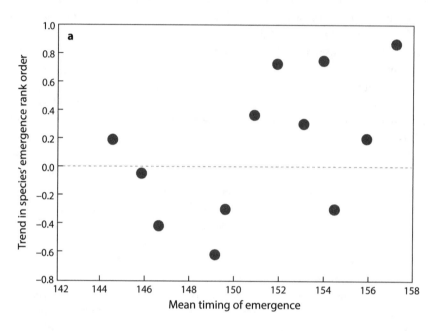

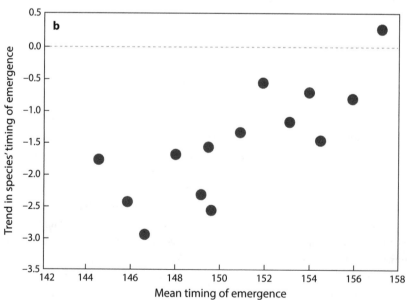

timing of emergence over the same period, a positive association appears (figure 5.4a). This association suggests that species with an early-onset life history strategy have tended to progress within the phenological community toward earlier ranks in the order of emergence. In contrast, species with a late-onset life history strategy have tended toward later ranks in the order of emergence within the phenological community. Moreover, the absolute trend in the annual timing of emergence by each species scales positively with the species' mean timing of emergence over the study period (figure 5.4b). This association suggests that species with an early-onset life history strategy have tended to undergo the most rapid advances, or have displayed a capacity for greater phenological plasticity, in their emergence phenology compared to species with later-onset life history strategies.

Such a relationship has been reported numerous times for a variety of taxa, including butterflies and other insects (Shapiro 1973; Hegland et al. 2009; Diamond et al. 2011; Ovaskainen et al. 2013), birds (Ovaskainen et al. 2013), reptiles and amphibians (Ovaskainen et al. 2013), and plants (Sherry et al. 2007; Munguia-Rosas et al. 2011; Sparks et al. 2011; Iler et al. 2013a; Ernakovich et al. 2014; Reyes-Fox et al. 2014; Wolkovich et al. 2014a; Post et al. 2016). The apparent generality of this pattern suggests that species may tend to segregate in time if abiotic or biotic constraints on the timing of expression of life history traits are alleviated. But does this imply that pockets of "empty time" will result? If some species advance their phenologies while others with which they co-occur in the local assemblage do not advance, or even delay, their phenologies, and if, furthermore, there is variation among those species that do advance or delay their phenologies in the extent to which they do so, might there arise unutilized temporal resources?

FIGURE 5.4. Associations between the life history strategies of individual species within the plant community at the Kangerlussuaq study site and (a) trends in their rank order of emergence ($r = 0.50$, $P = 0.10$) and (b) trends in their timing of emergence ($r = 0.79$, $P < 0.001$). Species' individual life history strategies are inferred in this example on the basis of their long-term mean timing of emergence derived from annual dates of initiation of aboveground growth from 2002 through 2013. Early life history strategists are those species with early mean emergence dates over this period, while late life history strategists are those with late mean emergence dates. The association in panel (a) suggests that early life history strategists have tended to undergo trends toward progressively lower ranks in the order of emergence of species, while the opposite applies to late life history strategists. The association in panel (b) suggests that this is due, at least in part, to stronger trends toward earlier timing of emergence in early life history strategists than in late life history strategists. Hence, early species have advanced their emergence phenology to a greater extent than have late species, resulting in a reshuffling of the phenological community at the study site since 2002. Adapted from Post et al. (2016).

NICHE PACKING AND CHANGES IN THE
AVAILABILITY OF ECOLOGICAL TIME

Evolutionarily, there should not exist in nature such a thing as a vacant niche, or unutilized niche space (MacArthur 1957; MacArthur and Levins 1964). Theory suggests niche packing should result from the processes filtering the local species assemblage from the regional species pool (Ricklefs 1987), and that niche packing should contribute to the maintenance of species co-existence in the local community (MacArthur and Levins 1967; May and MacArthur 1972; Ricklefs 2011). Partitioning of resources along ever-finer scales should result from colonization of local communities by new species, absent displacement of established species, and simultaneously promote coexistence (MacArthur and Levins 1964; Levine and D'Antonio 1999). By contrast, local extinction should promote expansion along niche axes partitioned through competition by remaining species, and utilization of forms of resources previously rendered unavailable through competition (Brown and Munger 1985). Do the same principles apply to the partitioning of time as a resource within the phenological community?

We should look for evidence of expansion or contraction along phenological niche axes, and of expanded or compressed use of time, in relation to increasing or decreasing availability of ecological time associated with changes in the abiotic or biotic environment. One line of evidence consistent with increased use of time or expansion into increasingly available ecological time with climatic warming might derive from studies documenting shifts in voltinism among invertebrates and other ectothermic organisms associated with recent climate change. In such organisms, annual periods of activity and metabolic- and developmental rates are sensitive to warming.

For instance, laboratory work demonstrates that the rate of development through some life stages of nematode parasites with indirect life cycles accelerates with warming (Kutz et al. 2001). In the Arctic, climatic conditions have recently become favorable to the completion of the life cycle of some parasitic nematodes in one summer season as opposed to two (Kutz et al. 2013). In some species of Heteroptera, or true bugs, physiologically based modeling indicates that warming on the order of 2°C to 3°C may increase voltinism by one generation per year (Musolin 2007). Similarly, voltinism in spruce bark beetles (*Ips typographus*) is expected to increase from one to two generations per year in southern Sweden if temperatures during the annual period of activity of the beetle undergo a comparable range of increase, from 2.4°C to 3.8°C (Jonsson et al. 2009). As well, a phenology model developed for the grape berry moth (*Paralobesia viteana*) that included the constraints placed by photoperiod on timing of diapause induction and termination suggests an increase in voltinism in this species with warming in excess of 2°C (Tobin et al. 2008).

As for empirical support for shifts in voltinism in field settings, a major review of studies of patterns of voltinism in 263 multivoltine species of butterflies and moths in European populations revealed evidence of earlier flight activity in 25 percent of species, and of an additional generation per year in 44 species (Altermatt 2010). Moreover, a comparison of the current timing of the annual flight season of the mountain pine beetle (*Dendroctonus ponderosae*) on the Colorado Front Range to historical data indicates that the species has increased its annual season of activity by one month (Mitton and Ferrenberg 2012). This was achieved through earlier onset of annual activity in association with climatic warming over the past two decades (Mitton and Ferrenberg 2012). According to Mitton and Ferrenberg (2012), concurrently, the species, which does not undergo photoperiod-driven diapause, has apparently undergone a shift in voltinism from one to two generations per year. However, that conclusion has been subsequently challenged on the basis that lower thermal thresholds that would constrain development through a winter generation at one site in the study by Mitton and Ferrenberg (2012) have not warmed to the extent necessary for existing constraints on a shift to bivoltinism to occur (Bentz and Powell 2014).

Exploitation of increased availability of ecological time in association with climatic warming should not be limited to ectothermic animals. Additional evidence consistent with this hypothesis might also be expected in shifts in the phenology of birds and plants. Among birds, such evidence might take the form of double clutching, or the production of more than one brood of hatchlings in a single reproductive season in individuals, populations, or species previously having produced only a single brood. A 17-year study of brood production by prothonotary warblers (*Protonotaria citrea*) in Virginia concluded that older females in the study population were more likely to engage in double brooding than were younger females, and they were more likely to do so during warm than during cold springs (Bulluck et al. 2013). Presumably, this owes to the fact that older females tend to arrive at breeding areas earlier than do younger females (Bulluck et al. 2013). As well, in warm springs when resource phenology is advanced, older females are more likely to capitalize on an earlier and truncated period of resource availability than are later arriving, younger females (Bulluck et al. 2013). Similarly, in a Virginia population of tree swallows (*Tachycineta bicolor*), a species that does not typically produce more than a single brood per year, the rate of double brooding during a 3-year study was substantially higher among females that initiated nesting before the population-wide peak timing of nesting (Monroe et al. 2008).

In contrast to such patterns, however, long-term studies (up to 50 years) of nesting phenology and offspring production in four populations of great tits (*Parus major*) in the Netherlands revealed declines in the production of second broods in all four populations in association with recent warming (Husby et al. 2009). Great tits engage facultatively in multiple breeding attempts per year, so variation or trends in the degree of double brooding in populations of this species may relate to changes in environmental conditions. In part, the decline in the proportion of females nesting in these populations that produced second broods annually was density dependent (Husby et al. 2009). However, after accounting for this, a clear decline in double brooding with recent warming was still apparent, especially for the most recent 30 years of the study when warming has been most pronounced (Husby et al. 2009).

There is some indication that the decline in rate of double brooding in association with warming in the study by Husby et al. (2009) may actually be driven by a mismatch between the timing of reproduction by great tits and the timing of peak availability of forage resources—namely, caterpillars. For one of the study populations, the successful production of offspring from second clutches declined with increasing mismatch between the timing of second brooding and the timing of the peak of caterpillar abundance, which has advanced more rapidly than the nesting phenology of great tits in the Netherlands (Visser et al. 2003). Hence, double brooding, rather than reflecting a strategy of capitalizing on the increased availability of ecological time with warming may, in instances such as this, pose no adaptive value if advancing phenology of forage resources outpaces any advance in the timing of initiation of reproduction. The role of relative ecological time in such consumer-resource interactions will be examined in further detail in chapter 7.

The preceding examples suggest that species tend to shift the timing of expression of life history traits related to growth and reproduction in response to the alleviation of constraints on their expression. Phenological packing, the compression of variation in timing of expression of life history traits among species within a single season, occurs during abiotically constrained or cold springs. By contrast, phenological segregation, the increase in variation among species in their timing of expression of life history traits within a single season, tends to occur in during abiotically amenable or warm springs. In chapter 8, examples of extreme phenological segregation will be presented from some tropical systems in which temperatures are conducive to biological activity throughout the year. Phenological packing and segregation have the potential to influence competitive interactions among species (Todd et al. 2011) by determining the extent to which, for instance, their emergence, arrival, flowering, or egg-laying timing overlaps with co-occurring species within the phenological community. In the next chapter, we will examine the implications of climate change for competitive, or interference, interactions in laterally structured phenological communities. In that and the following chapter, increasing emphasis will be placed on relative ecological time, the timing of phenological activity by an individual or species relative to the timing of phenological activity in those individuals or species with which it interacts.

CHAPTER SIX

Use of Time in the Phenology of Horizontal Species Interactions

Before examining the role of time in interactions among organisms, let us return briefly to the concepts of timing and duration as they relate to life history strategies of the individual that manifest as phenological dynamics. As we learned in chapter 2, rates and directions of phenological change vary considerably among species within traits. Among species of birds, for instance, the timing of arrival at breeding grounds has advanced at variable rates over the past several decades. While some species have undergone rapid shifts toward earlier timing of arrival, others have not changed at all. As well, the timing of initiation of spring growth or of flowering has advanced rapidly in some species of plants, while it has not changed or has even become delayed in other species. Moreover, within species, rates of phenological advance vary among traits. While timing of arrival at breeding grounds by some species of long-distance migratory birds, such as pied flycatchers (*Ficedula hypoleuca*), has not advanced, the timing of egg-laying by populations of the same species has (Both and Visser 2001, 2005). Similarly, the timing of successive phenophases in several species of plants has advanced at varying rates, altering interphase durations between, for example, emergence or bud break and flowering (Sparks et al. 2011). Furthermore, the most comprehensive analysis of phenological dynamics to date, encompassing over 10,000 unique data sets, documented remarkable variability within and among taxonomic groups in the direction and magnitude of response to climatic variability (Thackeray et al. 2016). Here, we will examine such disparities through the lens of adaptive variation in the expression of life history traits concerned with timing in the context of competition. In the following chapter, we will add to this interactions among species at adjacent trophic levels.

TIMING AND DURATION: RESPONSES TO POTENTIALLY DISTINCT SELECTION PRESSURES

To develop a more nuanced understanding of the selection pressures shaping the timing of expression of life history strategies related to growth or development, maintenance, survival, and reproductive success, first consider the simplest possible scenario involving the timing of life history events. This scenario concerns a unidimensional environment in which there are no competitors, no natural enemies, and no abiotic constraints on the timing of expression of life history events by the individual organism. Hence, it represents a near vacuum in which the individual organism contends solely with its own impulse to grow, survive, and reproduce successfully.

In this unidimensional environment, we may assume, if the timing of expression of any individual life history event is unconstrained, increasingly earlier expression of that trait will be advantageous to the individual if time is limited (*sensu* Williams 1966). Such advancement can be expected if the strategy thus favored by natural selection is an increase in the allocation of time to the succeeding life history stage. An advancement of the timing of expression of one life history event in the life cycle may also, however, abbreviate the duration of the preceding life history stage and so would not be favored by natural selection if the life history stage thus shortened were more critical to survival and offspring production than the ensuing life history stage. Under the former scenario, therefore, advancement of the timing of expression of life history traits will be favored only in situations in which it increases the allocation of time to development during a critical, ensuing life history stage. This scenario is depicted in figure 6.1a, in which the initial, or null, timing of expression of some trait influencing growth, survival, or reproductive success is represented by the solid black dot on the time axis. Deviation from this null timing in an unconstrained, unidimensional environment is represented by the arrows leading to progressively earlier expression of that life history trait in the progressively lighter gray dots as the availability of time increases (figure 6.1a).

Notably, however, this situation can be expected to unfold only in an environment free of constraints on the timing of expression of life history traits. Such constraints, when existent, may be abiotic or biotic in form. Examples of the former may include photoperiodic or solar irradiance limitation, temperature limitation, or limitation by moisture regime. Examples of the latter may include the presence of competitors of the same or different species, or of natural enemies whose presence limits the timing of expression of life history traits. Hence, we may impose simplistic constraints on the scenario in figure 6.1a by considering, first, the addition of competitors to the environment. To begin with, let us consider an environment containing only competitors of the same species. In such a scenario, the potential for intraspecific competition should select for increasingly early expression of life history traits concerned with timing, as depicted in figure 6.1b. This is because an environment in which the timing of activity is entirely identical and synchronous among all members of a population of a single species is an environment in which density-dependent competition for other resources may be expected to be disadvantageous to the individual (Iwasa and Levin 1995; Post et al. 2001b; Post 2013).

Hence, we may assume that the addition of a biotic element, in this case intra-specific competition, to the simple environment in figure 6.1a will act as a further driver of earlier timing of expression of life history traits concerned with growth, survival, and reproductive success. But there is an important contrast between the adaptive value of earlier timing in this instance compared to the simplest scenario in which competitors are lacking. In this instance, earlier timing is favorable to the individual because it presents an advantage in terms of competition for resources other than time, while in the competitor-free environment earlier timing presents an advantage because it promotes the allocation of more time itself to the duration of the life history stage that is initiated by the preceding life history event.

We may add further complexity to the environment by considering the additional influence of abiotic, environmental parameters such as temperature limitation in thermally seasonal environments. In this scenario, temperature limitation presents an abiotic constraint on earlier timing of expression of life history events, and this influence acts in opposition to the influence of competition (see figure 6.1b). Here, earlier timing, whether as a strategy aimed at the allocation of more time within the individual in a competitor-free environment, or whether as a strategy aimed at increasing access to other resources in a competitor-rich environment, will be hampered or rendered impossible until abiotic constraints are alleviated.

The result of the action of dual influences may be realized in a simple, two-dimensional environment consisting of a biotic promoter of earlier timing and an abiotic constraint on earlier timing (see figure 6.1c). In such an environment, the individual organism is forced to contend with the abiotic constraint on earlier timing while simultaneously responding to the biotic promoter of earlier timing. We can envision this environment as comprising two axes along which timing may be seen to progress in response to biotic and abiotic conditions (figure 6.1c). With timing in response to biotic promoters of advancement operating along the x-axis in figure 6.1c and timing in response to abiotic constraints on advancement operating along the y-axis, we can see that the rate of advance in timing of expression of a given life history trait along the x-axis in this environment is tempered compared to its rate of advance along the same axis in an environment comprising only intraspecific competition (figure 6.1b).

Hence, actualized advancement of the timing of expression of life history traits by the individual is a product of dual, and in this case opposing, selection factors. Expectations derived from such simplistic model environments should lead us to conclude that as abiotic constraints such as those imposed by temperature limitation become alleviated by climate change, earlier timing of expression of life history traits should ensue. But as has been repeated many times thus far, such is the case neither universally within species across traits, nor within traits across species. To understand this, we must consider the implications of timing for duration. In this context, duration denotes the length of time devoted to a single life history stage. When we consider a pair of life history events in a sequence, earlier timing of the first event increases the duration of the interval, or of the life history stage, following the first event. In such an instance, duration is important in an idealized, single-organism, competitor-free system because it determines life history progression through growth, survival, and reproduction. But in a multiorganism system, duration also determines phenological overlap among competitors. Thus, in a multispecies, competitor-rich system, duration additionally determines phenological overlap among competitors for other resources.

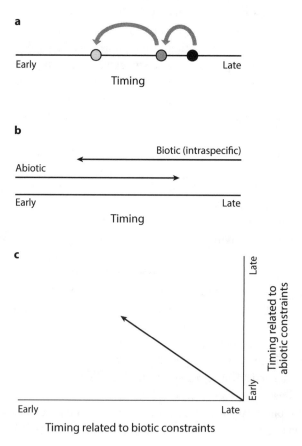

It follows that, while timing of expression of life history events is constrained by abiotic conditions and may also evolve in response to biotic factors involving competition and resource acquisition, duration may be seen as evolving in response to biotic factors related to competition for time within the individual and to competition for other resources among individuals of the same or other species. As we will explore in the following section, duration in the real world influences species overlap in time, and, thereby, the extent to which the individual experiences temporal crowding within the competitive phenological environment. I would suggest that selection on timing is, in actuality, selection on *duration* of critical phenophases determined by timing. If this is the case, then the discipline of phenology may benefit by increased focus on the implications of variation in timing for variation in duration.

From the scenarios outlined earlier, a notable insight emerges. The use of time by the individual must address potentially strongly competing interests. First, any variation in the timing of life history events may function, within the individual organism, as a mechanism for allocating as much time as possible to increase the

---

FIGURE 6.1. In panel (a), the tendency toward earlier timing of expression of a life history event may occur in a simplistic environment devoid of constraints. Here, the initial or default condition is defined by the relatively late timing of expression of a given life history event (the solid black dot) in a unidimensional environment comprising time only. In the absence of constraints on earlier timing of expression of this trait, earlier timing will be favored if the time allocated to progression through the next subsequent life history stage, or stages beyond this, scales positively with survival or reproductive success. Hence, progressively earlier expression of the life history trait (indicated by successively lighter gray dots), either in subsequent seasons by a long-lived organism, or in subsequent generations in a short-lived organism, will be favored. In panel (b), simplistic biotic and abiotic constraints on the timing of expression of a life history event are added to the environment. These may take the form of density-dependent competition arising from the presence of conspecific individuals in the former case, and of temperature, photoperiodic, or moisture limitation in the latter case. When both forms of constraint are in operation in the environment, they may exert opposing influences on the timing of expression of life history events by the individual. While it may be adaptive to become active or express life history traits related to growth, reproduction, and survival as early as possible to minimize exposure to competition in an increasingly crowded biotic environment as the active season progresses, abiotic constraints may impede phenological advancement. The resulting vector defining movement through time in a two-dimensional environment in which biotic and abiotic factors are in simultaneous operation is portrayed in panel (c). In this environment, progression through time is hindered by abiotic constraints operating along the vertical axis, while progression through time is favored by biotic promoters related to avoiding competition operating along the horizontal axis. Hence, the net result is a vector comprising elements characterized by differing rates of movement along both axes. By comparison to the simpler environment consisting only of biotic limitation, the rate of movement along the horizontal axis in panel (c) is reduced.

duration of critical life history stages. That same strategy may, however, expose the individual organism to increased competition for resources other than time. Therefore, individuals must balance the intrinsically opposing demands of increased allocation of time internally and reduced overlap in time with competitors externally. In this context, the search by ecologists for common patterns of phenological response to the alleviation of abiotic constraints on the timing of life history events by, for example, climatic warming, would seem misguided. The individual organism may become active earlier to serve internal demands related to increasing allocation of time to life history development through duration. In a competitive context, however, while earlier timing may in and of itself present a competitive advantage among conspecifics, its value as a strategy in interspecific competition relates to its effect on phenological duration. This is because duration determines overlap within the phenological community. Hence, increasing duration may be adaptive in a competitor-free environment but maladaptive or disadvantageous in a competitor-rich environment. In the latter case, earlier timing of life history events may, however, be an effective strategy for the use of time if it also advances duration between successive life history stages without increasing such duration.

VARIATION IN TIMING WITHIN AND AMONG INDIVIDUALS AND SPECIES

To understand the role of time in interactions among species within phenological communities wherein horizontal species interactions prevail, we must begin at the level of the individual within a species, and consider the allocation of time within its life history cycle. But first, a brief review of the concepts of horizontal and vertical species interactions is warranted. Horizontal, or lateral, species interactions are those involving individuals within a single trophic level in the same local community or species assemblage. These involve primarily interference interactions such as competition for resources required by more than one member of the local assemblage (Post 2013). Within the phenological community, the resource of main concern is time. Vertical species interactions, in contrast, are those involving individuals at different trophic levels within the same local community or species assemblage. These involve primarily exploitation interactions such as consumer-resource interactions involving predator and prey species, herbivore and plant species, pathogen and host species, or even mutualistic species (Post 2013). In the latter case, the exploitation interaction is bidirectional because each of the species involved in the mutualism exploits or depends upon the other as a resource in one form or another.

Beginning at the level of the individual, consider a highly simplistic life history schedule involving three events (figure 6.2). These might include, for

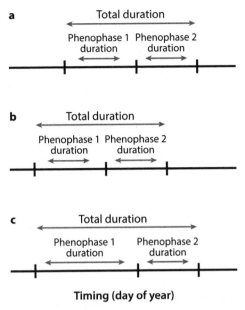

Events within an individual organism

Timing (day of year)

FIGURE 6.2. A simplistic depiction of the sequence of life history events, and the phenophases bounded by them, that constitute the life history of an individual organism. In this conceptualization, time proceeds within a season of activity along the horizontal axis in each panel from left to right, while time proceeds from one season to the next along the vertical axis from top to bottom. In panel (a), the three life history events are assumed to represent the timing of emergence, flowering, and fruit set in an individual perennial plant. Hence, phenophase 1 is the state of emergence, and phenophase 2 is the state of flowering. The durations of phenophase 1 and 2, defined as the number of days between emergence and flowering and between flowering and fruit set, respectively, are equal in this example, while the total duration of the life history is defined as the number of days from emergence to fruit set. In panel (b), the individual plant has advanced its timing of emergence, flowering, and fruit set relative to the timing of each event in the preceding season (panel a). Moreover, it has advanced the timing of each event equally relative to its timing in the previous season. As a result, neither the total duration of the individual's life history, nor the durations of either phenophase, has been altered. In panel (c), the individual has advanced the timing of emergence relative to its timing in the first growing season, while the timing of flowering and fruit set remain constant relative to the timing of each in the first growing season. In this instance, the total duration of the life history has increased owing to an increase in the duration of phenophase 1.

instance, the annual timing of emergence, flowering, and fruit set in an individual perennial plant. In this scenario, the timing of all three life history events is fairly evenly distributed among the dates comprising the annual season of productivity (figure 6.2a). The number of days between the timing of initiation of the first stage, or phenophase, in the life history cycle and the last phenophase constitutes the total duration of phenological activity between emergence and reproduction for this individual. By comparison, the number of days between the first and second phenophases, and between the second and third phenophases, constitutes the duration of phenophase 1 and the duration of phenophase 2, respectively (figure 6.2a).

As emphasized in the previous section, the timing of expression of one or more life history traits, and thereby the initiation of one or more phenophases, within the individual may be constrained by abiotic factors. If so, alleviation of such constraints should promote advancement of the timing of initiation of any such constrained phenophases. In panel (b) of figure 6.2, the timing of all three phenophases has advanced at equal rates in response to, for example, climatic warming. Hence, the entire life history schedule has advanced, and, while the durations of phenophase 1 and 2 have not been altered, more time may be allocated to, for instance, development of the seed or the seedling itself. Even though the timing of the entire life history schedule has advanced in relation to its timing in figure 6.2b, the total duration of the life history schedule remains unchanged.

In panel (c) of figure 6.2, only the timing of initiation of phenophase 1 has advanced, and its duration has increased. In contrast, the timing of initiation of phenophases 2 and 3 has remained fixed relative to their timing in the null scenario depicted in panel (a) of figure 6.2. Similarly, there has been no change in the duration of phenophase 2. Because only the timing of phenophase 1 has advanced, the total duration of the life history schedule in this case has increased relative to that depicted in the scenario in panel (a) of figure 6.2. A corresponding increase in the total duration of the life history schedule would ensue from a delay in the timing of onset of phenophase 3 if the timing of initiation of phenophases 1 and 2 remained fixed.

Admittedly, this is a simplistic scenario, but we can employ it to illustrate the manner in which individuals within a species may respond to alleviation of abiotic environmental constraints on the timing of expression of individual life history traits within the total life history schedule or cycle. Such phenological shifts within the individual have the potential to alter the extent to which the individual overlaps in time with other members of the same species, and hence the degree of density-dependent, intraspecific competition experienced by the individual. We may expect, therefore, the durations of phenophases within the individual and of the total life history schedule to *increase* in response to positive selection on earlier timing aimed at reducing intraspecific competition, but to *decrease* in response to negative selection owing to intraspecific competition itself. The dual and opposing selection pressures imposed by competition on timing and duration of phenophases and the total life history schedule may be expected to result in segregation in time among individuals of the same species.

The null scenario on the left side of figure 6.3 depicts the distribution of conspecific individuals according to the timing of expression of, for simplicity, a single life history trait. In this example, let us assume this panel represents the timing of emergence by conspecific individuals of a single species of flowering plant. The horizontal axis in this panel represents time or day of year during the growing season, and the white dots represent the dates on which individuals of this species have emerged in year t. Above the individual dates is a box-and-whiskers plot depicting the median date of emergence for the population, and the 25th and 75th percentiles and endpoints of the distribution of emergence dates. Hence, there is a clustering of individuals around the median, and a reasonably normal distribution of emergence dates around that, with single outliers at each end. Such variation forms the basis for selection on the timing of expression of this life history trait in response to variation or changes in abiotic environmental constraints and competition.

If we assume the timing of emergence in this species is constrained by springtime temperature, then the three panels (a, b, and c) on the right side of figure 6.3 depict potential changes in the species-level pattern of expression of this life history trait in response to climatic warming. But examples of limitation by any abiotic factor could just as well be used to illustrate the following point. First, in panel (a), warming has elicited an equivalent shift in the timing of emergence by all individuals within this population of the species. Hence, while the median date of emergence has advanced, the distribution of individuals around the median is unchanged from that depicted in the null scenario. Alternatively, panel (b) of figure 6.3 illustrates an asymmetrical advance in the timing of emergence by the earliest emerging individuals in the population compared to middle- and later emerging individuals. In this instance, the median and later dates of emergence within the population remain unchanged relative to the null scenario, but the species-specific timing of initiation of emergence has advanced. Consequently, at the level of the species, the total duration of emergence has increased, even though the median date of emergence has not changed. As the next section will emphasize, the same dynamic should apply at the community level to differences among species.

Last, in the third scenario, depicted in panel (c) of figure 6.3, the entire distribution of emergence dates within this species has shifted toward earlier emergence with warming, but it has done so to a greater extent for early- than for late-emerging individuals. Hence, the species-level duration of the timing of emergence has increased simultaneously with an advance in phenology toward earlier emergence by the species as a whole. As we will see in the next section, these three scenarios, which depict changes in the "phenological behavior" of individuals within a species, pose varying consequences for the extent of overlap in time with other species in the phenological community. Such consequences

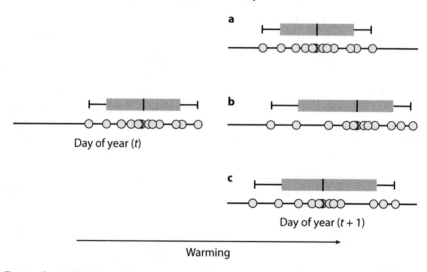

FIGURE 6.3. A depiction of the manner in which responses by individuals to the alleviation of abiotic environmental constraints on the timing of expression of a life history trait can result in changes in patterns of variation in the timing of expression of that trait within a population at the species level. Here, the timing of expression of the life history trait of interest is assumed to be temperature limited; hence, alleviation of the operative constraint results from climatic warming. Time within each season or year proceeds along the horizontal axis in each panel from left (early) to right (late), and likewise between years from left to right. In each panel, the vertical lines within box plots represent the mean or median timing of expression of the life history trait at the population level, while the left and right ends of the box plots represent the 25th and 75th percentiles of the distribution of dates of expression, and the whiskers represent the earliest and latest dates of expression in the population. Gray dots along the time axis in each panel represent the timing of expression of the life history trait by each individual in the population. Three possible scenarios can result in the transition from season or year one, on the left, to season or year two on the right. First (a), all individuals of the species may advance the timing of expression of the life history trait equally. As a result, the entire distribution of dates of expression at the population level advances equally. Second (b), individuals characterized by early expression of the trait in year one advance their timing to a greater extent than do individuals characterized by late timing in year one, and the latest individuals do not advance at all. As a result, the population-level distribution of dates of expression becomes left-shifted, but the mean or median may remain constant. Importantly, individuals, especially those at the early end of the spectrum of dates of expression, have segregated in time and, presumably, experience reduced competition in time as a result. Last (c), all individuals have advanced their timing of expression, but early individuals have done so to a greater extent than have late individuals. As a result, the entire distribution of dates of expression of the life history trait has shifted to the left, and all or most individuals in the population have segregated from one another in time, presumably reducing intraspecific competition in time.

emerge from the scaling up of individual life history strategies to the level of the species, and from there to the level of the local species assemblage or community. This scaling up is a consequence of the application of the same selection pressures occurring at the individual level to dynamics seen at the species level within the community.

VARIATION IN TIMING AMONG SPECIES WITHIN COMMUNITIES

If individuals of a given species segregate in time in the expression of life history traits, what are the implications of such segregation for overlap in time with individual members of potentially competing species? In figure 6.4a, we see a depiction of the timing of expression of a single life history event by members of a simple, three species community. The box-and-whiskers plots in panel (a) of figure 6.4 represent the median, 25th and 75th percentiles, and endpoints of the dates of emergence for each of the three species in this local assemblage. These are comparable to the distribution of emergence dates for the hypothetical single species depicted in figure 6.3. Hence, in this null scenario, emergence occurs first, on average, in species 1, followed by species 2, and, last, species 3. While species 1 and 2 display relatively similar distributions of emergence dates, with considerable variability around the median within the 25th and 75th percentiles, species 3 displays a comparably tighter distribution of emergence dates about the median with extreme outliers on either end of the 25th and 75th percentiles (figure 6.4a). We can surmise from the shapes of these distributions that most individuals of species 3 potentially experience considerably stronger intraspecific than interspecific competition in time, while individuals of species 1 and 2 experience relatively comparable intra- and interspecific competition in time. However, species 2 is overlapped by both species 1 and species 3, and hence individuals of species 2 experience the strongest interspecific competition in time of the three species in this community.

From this null scenario, we can explore the implications for intra- and interspecific competition in time of the differential and species-specific phenological responses to alleviation of abiotic constraints on the timing of emergence. Warming from panel (a) to panel (b) in figure 6.4 has elicited phenological advances in the timing of emergence in both species 1 and species 2, but a stronger advance in species 1. Species 3, in contrast, has not advanced its emergence phenology in response to warming. None of the three species has undergone a change in the distribution of emergence dates about its median date of emergence. The result is a greater segregation of all three species in time, and hence a reduction in the degree of interspecific competition for all three species, without any alteration in the strength of intraspecific competition in time for any of the species. Notably,

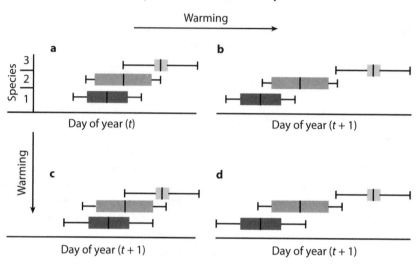

FIGURE 6.4. An example of a simple, hypothetical phenological community comprising three species at a single trophic level. Species-level patterns of variation in the timing of expression of life history traits arise from individual-level responses to the alleviation of an abiotic environmental constraint on trait expression. In this example, phenology is constrained by temperature, and individuals within species may respond to warming by advancing the timing of expression of a life history trait such as emergence. Panel (a) thus represents the distributions of emergence dates of individuals of each of three species in the current season or year. Box plots are representative of the population-level distributions of dates within each species, as shown in figure 6.2. Proceeding from panel (a), warming may result in species-specific changes in the mean or median date of emergence and variation about this date in the subsequent season or year that alters the phenological community in one of three possible ways. For instance, from panel (a) to panel (b), warming has elicited a shift toward earlier emergence in species 1 and 2, but a greater shift in species 1 than in species 2. Species 3 has not altered its timing of emergence. In none of the species has the distribution of emergence dates among individuals been altered by warming. As a result, the magnitude of interspecific competition in time has been reduced for all three species, but the magnitude of intraspecific competition remains unchanged in each species. From panel (a) to panel (c), warming has not elicited a shift in the mean or median timing of emergence in any of the three species, but it has resulted in an increase in the dispersion of dates about the mean in species 1 and 2, though to a greater extent in species 1. In this instance, the magnitude of interspecific competition in time experienced by all three species increases, while the magnitude of intraspecific competition among individuals is reduced in species 1 and 2. From panel (a) to panel (d), species 1 and 2 have advanced their mean or median timing of emergence while simultaneously increasing the dispersion of emergence dates about the mean or median, but to a greater extent in species 1 than in species 2. In this case, the earliest emerging individuals of species 1 enjoy reductions to both intra- and interspecific competition in time, while the same applies to the latest emerging individuals of species 2. Individuals of species 3 experience reduced interspecific competition in time relative to that experienced in the scenario depicted in panel (a).

species 3 has achieved this increase in segregation in time from species 1 and 2 without individuals of that species having altered their own timing of emergence.

In contrast, warming from panel (a) to panel (c) has resulted in no shift in the median date of emergence for any of the three species. Warming has, however, elicited an increase in the variation in emergence dates about the median date for both species 1 and species 2. Such an increase in the variation in species-specific emergence dates with warming in this scenario has resulted from the earliest individuals advancing their timing of emergence and the latest individuals delaying their timing of emergence in both species. Individuals of species 3 have not altered their timing of emergence. The result in this case is an alleviation of intraspecific competition in time within both species 1 and 2, but an intensification of interspecific competition in time for all three species, most notably for species 2.

Last, warming from panel (a) to panel (d) has resulted in the most complex mix of phenological shifts. The timing of emergence by species 3 remains, once again, unaltered by warming. However, both species 1 and 2 have advanced their timing of emergence while simultaneously undergoing an increase in the variability in emergence dates about the median date within each species. Compared to the null scenario, the compound effect of this dual response to warming is a coincident alleviation of intraspecific and interspecific competition for species 1 and 2, and an alleviation of interspecific competition for species 3.

EMPIRICAL APPLICATIONS TO HORIZONTAL PHENOLOGICAL COMMUNITIES

A description of ongoing research on plant phenology at the long-term study site near Kangerlussuaq was presented in *Ecology of Climate Change* (Post 2013), but a brief review will be given here. Phenophases of all species of plants occurring on 12 marked plots are recorded annually on a daily or near-daily basis beginning in early to mid-May and extending into late June or July. The observational plots are grouped at three study areas within the larger study site, and these are separated by approximately 1 to 1.5 km. The exact timing of the start and end of the period of observation varies from year to year depending somewhat on logistical constraints. Because of this variation, the most useful data deriving from this effort are data on species-specific timing of emergence, or onset of activity, and on community-level median emergence.

A plot of the mean dates of emergence by all species observed each year reveals considerable variation among years in the timing of emergence at the community level; also notable is variation in the dispersion or scatter of species' mean annual

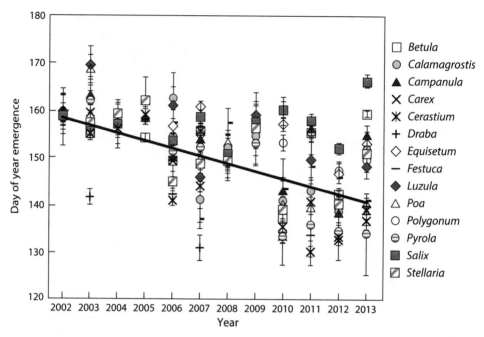

FIGURE 6.5. Variation in the plant phenological community, both among individuals within species and among species, over 12 years of observation at the Kangerlussuaq study site, from 2002 through 2013. In each year, the mean (±1 SE) date of emergence is shown for each species observed in that year. The black trend line is fit to data on the annual dates on which 50 percent of the species observed that year had emerged. The trend is significant ($R^2 = 0.56$, $P = 0.005$), and indicates a rate of advance of the midpoint of community-level emergence of approximately 16 days per decade (Post et al. 2016).

emergence dates within each year (figure 6.5). First, take note of the trend in community-level median emergence date. This is indicated by the solid black trend line in figure 6.5. Community-level median emergence is the date on which half of the species present on the plots in any given year have emerged. This date has advanced by approximately sixteen days per decade since observations began in 2002 (Post et al. 2016).

Next, note that in some years, the degree of species overlap in time is markedly pronounced. This is especially evident in the years 2002, 2004, 2005, 2008, and 2009. All of these years tend to have undergone later than average emergence. As noted earlier, the years 2004 and 2005 experienced a severe outbreak of the larvae of the noctuid moth, *Eurois occulta*. This outbreak reduced not only the numbers of species observed on phenology monitoring plots in those years, but also may have constrained emergence times of the remaining species.

Additionally, the spring seasons of 2002 and 2009 were among the coldest at the study site thus far.

In contrast, in years when emergence is early, on average, species' individual emergence dates are more greatly dispersed in time (see figure 6.5). This suggests, as discussed in the previous chapter, that some species tend to capitalize on increased availability of ecological time with the alleviation of abiotic environmental constraints on the timing of expression of life history traits. In fact, in this example, species with early-emergence life history strategies, such as *Pyrola* sp., *Cerastium* sp., and *Draba* sp., tend to advance their emergence phenology more rapidly than species with late-emergence strategies such as *Salix* sp., which has not advanced at all (Post et al. 2016). As a consequence, species become more temporally segregated in early years, as in the hypothetical warming scenario presented earlier in figure 6.4b. As well, there appears to be some indication, based on examination of the error bars about the means in figure 6.5, that early emergence at the species level is characterized by greater species-specific variability in the timing of emergence. A test of this hypothesis will be presented later in the chapter. This may reflect greater compression in time of emergence dates during cooler or later-emergence years as individuals within species become squeezed into more highly compressed windows of favorable time compared to warmer or early-emergence years. If so, the same pattern should be evident with variation among species at the community level.

In fact, there is a negative association between variation among species in their timing of emergence and mean community-level emergence (figure 6.6a; $r = -0.69$, $P = 0.01$). Here, variation is estimated as the coefficient of variation (CV) of mean emergence dates across all species in a given year. Hence, each point in figure 6.6a is based on estimates of variability in emergence dates and mean emergence dates for all species combined within a single year. This allows us to examine changes in the dispersion of species-specific emergence dates in relation to changes in community-level emergence. The association in figure 6.6a accords with the notion that years of early overall emergence at the community level are characterized by greater segregation among species in their individual dates of emergence.

Next, we may examine the same association within species, rather than at the community level (figure 6.6b). In figure 6.6b, each point represents a species. For each species, we see variation in its timing of emergence among years plotted against its mean date of emergence for the period of observation (2002–2013). Here, again, earlier emergence is associated with greater interannual variability in timing of emergence ($r = -0.86$, $P < 0.001$). In other words, species with an early-emergence life history strategy, such as the flowering forbs *Pyrola* sp., *Cerastium*

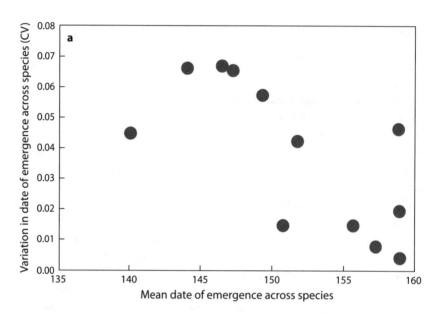

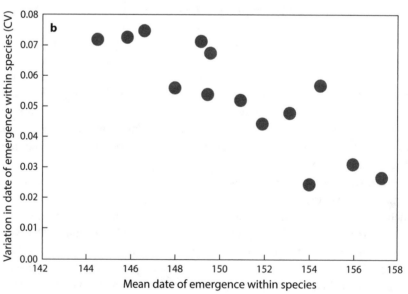

sp., and *Draba* sp., tend to display greater variability among years in their timing of emergence. In contrast, species with late-emergence life history strategies, such as the dwarf shrub *Salix* sp., tend to be comparatively stable from year to year in their timing of emergence.

Evidence from experimental studies indicates similar shifts in species-specific phenology and changes in community-level phenological dynamics with warming. A one-year warming experiment was conducted in tallgrass prairie at a field site in Oklahoma using overhanging infrared heaters (Sherry et al. 2007). The budding, flowering, and fruiting phenology of the 12 most dominant species of plants at the site, which accounted for approximately 90 percent of community-level biomass, were monitored for changes in timing of all three phenophases, and duration of the entire reproductive phase, in response to the warming treatment. Warming advanced the flowering phenology of 8 of 9 species with early season life history strategies, while it delayed the flowering phenology of 2 of 3 species with late-season life history strategies (figure 6.7) (Sherry et al. 2007). These shifts were matched by advances in fruiting dates of the same 8 of 9 early-season species. Two of the three late-season species delayed fruiting in response to the warming treatment, but only one of these had also undergone delayed flowering in response to warming. As a consequence of shifts in the timing of flowering and fruiting, the duration of the total reproductive phase was altered by warming in 7 of the 12 species (Sherry et al. 2007). The most notable result of this experiment was, however, the increase in temporal segregation between early and late life history strategists at the community level in response to warming (figure 6.7).

FIGURE 6.6. Community-level and species-specific patterns of variation in the dispersion of emergence dates with emergence timing based on observational data from the plant community at the Kangerlussuaq study site from 2002 through 2013. In panel (a), community-level variation among species in emergence timing increases with earlier emergence at the community level. Hence, species appear to segregate more in time in conjunction with earlier emergence, presumably reducing interspecific competition in time in response to the alleviation of abiotic environmental constraints on emergence timing. This apparently results from differences among species in their rates of phenological advance because early species advance at greater rates than late species as environmental constraints diminish with warming (Post et al. 2016). In panel (b), variation among individuals within species in the timing of emergence increases with earlier species-specific timing of emergence. In this case, individuals within species appear to segregate more in time with earlier emergence. Presumably, this pattern relates as well to differences in rates of advance among individuals within species, with early individuals advancing more rapidly than late individuals. However, this hypothesis has not been tested directly with the Kangerlussuaq data (Post et al. 2016). Together, the patterns in panels (a) and (b) appear consistent with the scenario depicted in the progression with warming from panels (a) through (d) in figure 6.4.

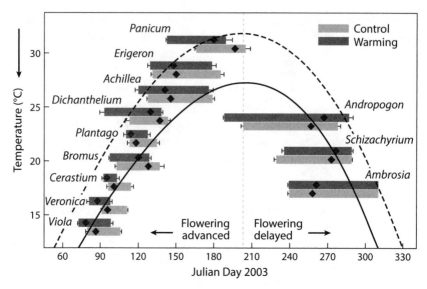

FIGURE 6.7. Alteration of the reproductive phenology of an entire plant community by experimental warming of a tallgrass prairie during a single growing season (March–November) at a field site in Oklahoma. Evident are divergent phenological responses by early- versus late-season life history strategists. For each of the 12 species, the upper bars represent mean (±1 SE) durations of the entire reproductive season on warmed plots, while lower bars represent mean durations on control (unwarmed) plots. The reproductive season for each species comprises the dates of budding, flowering, and fruiting. The dashed curve represents a polynomial regression line fit to daily mean temperatures on warmed plots, while the solid line represents a polynomial regression fit to daily mean temperatures on control plots. The dashed vertical line is the date on which maximum daily temperature occurred, and which delineates the distinction between early- versus late-season life history strategists. Diamonds within bars are the mean first flowering dates by treatment for each species. Warming significantly advanced the timing of the reproductive season for early species and delayed it for late species. Moreover, species displayed highly individualistic alterations of their total reproductive phenology in response to warming, in some cases reducing it and in others extending it, with consequent implications for species overlap in time. Modified from Sherry et al. (2007).

Hence, like the observational study of species-specific and community-level plant phenological dynamics at the Kangerlussuaq study site, this example appears to comport with the hypothetical example in figure 6.4b.

INDICES OF PHENOLOGICAL OVERLAP AMONG SPECIES

The preceding examples drew associations between alterations to species-specific timing and duration of phenophases and the alleviation of abiotic environmental constraints on the expression of life history traits associated with timing. Both of these case studies, deriving from widely differing types of plant communities, suggest that advances in the timing of emergence or flowering result in increased segregation among species within the phenological community. In both systems, this resulted from early-season life history strategists advancing the timing of expression of life history traits to a greater degree than late-season life history strategists, or in combination with delayed expression in late-season life history strategists. However, these examples were largely descriptive of changes in overlap among species within the phenological community.

Quantitative approaches to the analysis of phenological overlap among species co-occurring within the local assemblage have a fairly extensive history in ecology, and relate to early methods of quantitative assessment of niche overlap (Fleming and Partridge 1984; Rafferty et al. 2013). Historically, niche overlap has been quantified by employing an index of pairwise similarity between species in their use of, or position along, some niche axis (Schoener 1970; Pleasants 1980). In the study of phenological interactions or associations within the local community, the metric of interest is an index of overlap in the occurrence of one or more phenophases such as, in plants, the timing of emergence or flowering (Fleming and Partridge 1984; Rafferty et al. 2013). Pairwise indices quantify phenological overlap between pairs of species, while some composite of all possible pairwise indices provides a metric of community-wide phenological overlap. Commonly, the pairwise index, when modified for use in the quantification of phenological overlap, takes the form:

$$\alpha_{ij} = 1 - \frac{1}{2}\sum_{k=1}^{n} |p_{ik} - p_{jk}|, \tag{6.1}$$

in which p_i and p_j represent the proportions of the total number of individuals of species i and j that are observed to be in a given phenophase at time k (Schoener 1970; Fleming and Partridge 1984; Kraft and Ackerly 2010; Dante et al. 2013). This index is also known, in the niche literature, as a proportional similarity measure (Slobodchikoff and Schulz 1980). The index in equation (6.1) may then be averaged for all species pairs co-occurring in the local assemblage to derive an index of community-level mean phenological overlap for a given date or for the season as a whole (Dante et al. 2013). Such an index, both when calculated for pairwise species-level overlap and for community-level mean phenological overlap, may be compared for treatment versus control means in experimental studies, or analyzed through time in observational studies, to investigate changes in overlap within and across the phenological community. It can also be employed to compare the mean pairwise overlap for individual species within the community to the community-level mean overlap. This provides a basis for inferring the degree to which individual species experience more or less phenological overlap than the community on average, and whether and to what extent this is altered by the alleviation of abiotic environmental constraints on the timing of life history events. The inverse of such overlap among competing species may be interpreted as an index of the availability of ecological time. In the next chapter, when we consider overlap from the perspective of a consumer, it is a direct index of the availability of ecological time.

Yet another index of community-level phenological overlap relates to "intensity" (Diez et al. 2012). In this approach, the number of species engaged in a given phenophase on a given date is summed across the entire community (Dante et al. 2013). This total then provides, for any date within the season, an index of the intensity of engagement in that phenophase for the phenological community, a potential index of the intensity of competition along the temporal niche axis within the local assemblage.

While the proportional similarity measure represented by equation (6.1) and the community intensity index provide measures of the aggregation or dispersion of species in time, they do not quantify variation within species or thereby intraspecific aggregation or dispersion in time. Simultaneous measures of inter- and intraspecific aggregation or dispersion are necessary for investigating the manner in which species-specific responses to the alleviation of constraints on the timing of life history events ripple through or attenuate across the phenological community. Such is the amplification of phenological dynamics from individuals within species to species within communities conveyed from figures 6.3 to 6.4.

Returning briefly to figure 6.5, we find time series of the mean dates of emergence recorded for all species observed in each year of the ongoing monitoring study at Kangerlussuaq. In this plot, it appears, based on a cursory examination of the species-specific mean dates of emergence in each year, that there is increased segregation among species in time as the mean date of emergence across species advances. In other words, during years of early average timing of emergence, when abiotic constraints on species-specific emergence timing are presumably alleviated, species tend to segregate temporally. However, looking closer, it also appears that the standard errors about the species-specific means in each year also increase with advancement of the timing of emergence. If so, this would suggest that individuals within species also tend to segregate temporally with earlier initiation of emergence during years when abiotic constraints on emergence are alleviated. Such dynamics would appear to be representative of the type of phenological shift represented in figure 6.4 from panel (a) to panel (d): increasing segregation among the species-specific mean dates of a phenophase accompanied by increasing intraspecific variation about the species-specific mean dates. Development of the means to test this hypothesis quantitatively requires, therefore, simultaneous indices of dispersion among individuals within species and pairwise mean differences in timing of phenophases between species within the community.

To derive a metric of dispersion among individuals within species that can be expressed for the entire community, the species-specific range of dates over which a particular phenophase is initiated may be compared to the range of dates over which the same phenophase is initiated by any or all other species in the local assemblage. This comparison yields an index of overlap in time between pairs of species that can be averaged over the entire local assemblage. This provides an index of community-level phenological overlap for the phenophase of concern. For any pairwise comparison within the community, this index may take the form:

$$P_{t(i,j)} = D_{U(MINi,j)} - D_{L(MAXi,j)} \tag{6.2}$$

in which $P_{t(i,j)}$ is the index of phenological overlap for a phenophase observed in species i and j in year t, $D_{U(MINi,j)}$ is the lesser of the two dates of the upper range limit of the phenophase for the species pair i,j, and $D_{L(MAXi,j)}$ is the greater of the two dates of the lower range limit of the phenophase for the species pair i,j. The average of all of the pairwise differences thus provides the community-level mean overlap for the phenophase of interest in year t.

Deriving a metric of mean pairwise differences in timing of a phenophase that can be applied across the local species assemblage requires simply calculating the absolute value of the difference in timing of a phenophase between pairs of species, and then applying this to all possible pairwise species combinations. The mean of this value for all pairwise mean differences provides a community-level metric of dispersion among species-specific mean phenophase dates. This community-level mean pairwise phenophase difference can then be calculated for each year of data.

To illustrate this approach, return to the data in figure 6.5. First, we want to test the hypothesis that segregation among species in their mean timing increases with earlier overall timing of expression of the phenophase of interest—in this case, the timing of emergence. This is the scenario representing increased segregation of mean dates among species owing to differential rates of phenological advance among early- versus late-season strategists in figure 6.4. If early-season specialists advance their timing of expression of life history traits to a greater extent than late-season specialists, we should expect increasing segregation of species along the time axis when environmental constraints on early-season activity are alleviated. Second, we want to test the hypothesis that dispersion among individuals within species has increased with earlier timing of emergence. Such would be the result if phenological dynamics among species were reflective of life history dynamics within species. That is, we may expect greater dispersion of the timing of life history events within species with earlier occurrence if, within species, early-season individuals advance their timing to a greater extent than late-season individuals. This scenario, depicted in panel (d) of figure 6.4, should also, perhaps counterintuitively, result in greater overlap among species at the community level despite increased segregation among species' mean dates of emergence. Hence, testing this hypothesis requires two components: examining the relationship between species-specific dispersion (that is, ranges of emergence dates among individuals) and emergence timing, and examining the relationship between community-level overlap in species-specific ranges of emergence dates and community-level timing of emergence.

To test the first hypothesis, we must first calculate all pairwise mean differences in emergence dates among species for each year of data in figure 6.5, from 2002 through 2013. Next, we plot these against indices of the timing of community emergence. Previously, we derived a means of estimating the timing of community-level onset of emergence and its progression based on the proportion of the total number of species present on our monitoring plots (Post et al. 2003). This index quantifies the daily proportion of all species observed in a given phenophase, such as emergence, and is calculated as:

$$\Phi = \frac{1}{1 + e^{-(a + bX)}} \tag{6.3}$$

in which Φ is the proportion of species observed in the phenophase of interest, and X is the day of observation. Equation (6.3) can thus be utilized to estimate the date on which any proportion of species is present in any given phenophase. Hence, this index provides a measure of the timing of onset and progression of any phenophase of interest at the community level. If we want to estimate the date on which 5 percent of species are emergent each year, an index of the timing of onset of community-level emergence, we can solve equation (6.3) for $\Phi = 0.05$ (Post and Klein 1999; Post et al. 2003, 2008b; Kerby and Post 2013b).

In panels (a) and (b) of figure 6.8, we see the index of community-level mean pairwise differences among species in their annual timing of emergence plotted against two indices calculated using equation (6.3) for the onset (a) and midpoint (b) of community-level emergence. Here, the timing of onset of community emergence is defined, as described earlier, as the annual date on which 5 percent of species have emerged, while the timing of the midpoint of community emergence is defined as the date on which 50 percent of species have emerged. Hence, the plots in panels (a) and (b) of figure 6.8 depict the degree of overlap among species in their mean dates of emergence in relation to the annual timing of the start and middle of the growing season. Both associations are negative, but only that associated with the onset of community emergence is significant ($r = -0.58$, $P = 0.05$). This indicates that, indeed, seasons characterized by an earlier start to the growing season—that is, by more advanced emergence phenology—are also characterized by greater segregation in time among species in their mean dates of emergence.

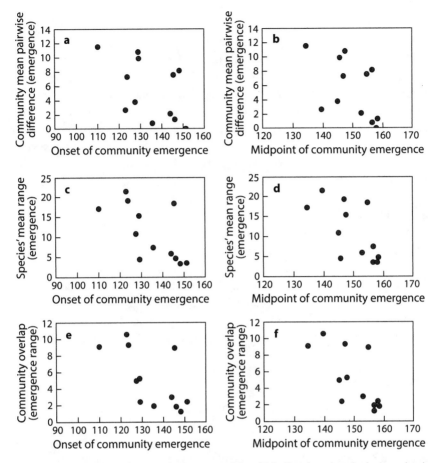

FIGURE 6.8. Tests of whether, to what extent, and in which direction phenological overlap in time among and within species varies with the timing of emergence within the plant community at the low-Arctic field site near Kangerlussuaq, Greenland. In this example, using 12 years of species-specific dates of emergence presented in figure 6.5, various metrics are employed to determine whether overlap in time increases or decreases with earlier emergence. In panels (a) and (b), mean pairwise differences in emergence dates among species increase with earlier community-level emergence, both for the onset (that is, start) of emergence (a) and the midpoint (b) of the emergence season. Hence, as the community becomes active earlier in spring, species' mean emergence dates appear to segregate in time to a greater extent than is the case during later years. In panels (c) and (d), the dispersion of emergence dates among individuals within species similarly increases with earlier onset of emergence at the community level and with an earlier midpoint of the emergence season. Hence, individuals within species appear to increase segregation in time with earlier onset. Presumably as a consequence of this, there is increased overlap among the ranges of species-specific emergence dates with earlier onset (e) and an earlier midpoint (f) of the season of emergence. This suggests that the earliest and latest individuals within species tend to experience simultaneously reduced temporal overlap with conspecific individuals in association with earlier emergence but increased temporal overlap with individuals of other species under the same conditions.

To test the second hypothesis, we must first test for an association between the mean species-specific range of emergence dates among individuals and the timing of community emergence. This can be achieved by, first, simply deriving the range of emergence dates for each species in each year of data depicted in figure 6.5, and, second, plotting these against the annual indices of the timing of onset and midpoint of community emergence derived as described earlier. Doing so reveals negative associations between the mean species-specific range of emergence dates and the timing of the onset and midpoint of community emergence (figures 6.8c and 6.8d, respectively). Both of these associations are significant ($r = -0.65$, $P = 0.02$; and $r = -0.64$, $P = 0.02$). Hence, species are, in fact, characterized by greater intraspecific dispersion of emergence dates in seasons of early emergence.

Next, we need to test for an association between the extent of overlap among species in their ranges of emergence dates and the timing of community emergence. To do this, we calculate $P_{t(i,j)}$ from equation (6.2) for all species pairs in each year covered by the data in figure 6.5. We then plot these against our indices of the annual timing of onset and midpoint of emergence. The resulting plots reveal negative associations that are significant in both cases, though with a slightly weaker correlation for the onset of community emergence (figure 6.8e; $r = -0.64$, $P = 0.02$) than for the midpoint of community emergence (figure 6.8f; $r = -0.69$, $P = 0.01$). Hence, we find evidence that, despite greater segregation among species in their mean dates of emergence during early-onset years, such years are also characterized by greater overlap among species in their ranges of emergence dates. This result accords with the scenario depicted in panel (d) of figure 6.4.

Because the patterns we have explored using the Kangerlussuaq phenology data to this point may be slightly challenging to dissect and hold clear in our minds, a brief review is warranted. First, earlier emergence is associated with greater variability among and within species in the timing of emergence (figure 6.6). This suggests increased use of time with increasing availability of time, a pattern that apparently applies among species as well as within them. Second, differences among species in the mean timing of emergence increase with earlier emergence (figure 6.8a,b). This indicates increased segregation in time among species, on average, as availability of time increases. Similarly, individuals within species display increased dispersion with earlier emergence (figure 6.8c,d). This indicates increased segregation in time within species as availability of time increases, matching the pattern evident at the community level among species.

The concordance of these patterns suggests that differences among early- and late-emergence life history strategists in their rates of phenological advance relate to differences among individuals within species as well as to differences among species within communities. Hence, community-level phenological patterns relating to the use of time reflect species-specific patterns emerging from individual-based strategies.

Last, although species appear on average to increase segregation among one another with earlier emergence, the net effect of the same tendency toward increased temporal segregation within species results in greater overlap among the endpoints of emergence timing among species (figure 6.8e,f). This suggests that there is variation among individuals within species, as well as among species within communities, in the strategies employed in coping with the consequences of phenological advancement for changes in intra- versus interspecific competition in time. How do we make sense of these apparently opposing patterns?

We might find it fruitful in this context to examine the extent to which, first, the range of dates of emergence within a species correlates with that species' emergence timing. Next, we may then examine the extent to which such correlations, if they occur, scale with species-specific trends in emergence. Such an examination may provide further insight into the consequences for phenological community dynamics of intra- versus interspecific dispersion of the expression of life history traits related to timing. Figure 6.9 shows such a plot using the Kangerlussuaq emergence data. Along the y-axis, we see species-specific correlations between emergence timing and the range of emergence dates calculated across all years of data.

Most of these values are negative, indicating that earlier emergence is associated with a greater range of emergence dates within each species. But two species, the flowering forb *Pyrola grandiflora* and the dwarf shrub gray willow *Salix glauca*, display positive correlations, indicating reduced ranges of emergence dates with earlier emergence in both species. We can make sense of the apparently aberrant correlations in these two species when the values for all species are plotted against the species-specific trends in emergence across all years of data. The parabolic nature of this association (figure 6.9) reveals an interesting pattern. The two species with the strongest and weakest emergence trends are *Pyrola grandiflora* and *Salix glauca*, respectively. In fact, the trend for *Salix glauca* is slightly positive, but not significantly so. Hence, the species at each end of the spectrum of trends in emergence timing are those in which earlier emergence is accompanied by reduced dispersion, or greater temporal compression, in emergence dates among individuals. The species undergoing the greatest rate of advance in emergence timing in this community, *Pyrola grandiflora*, apparently segregates itself temporally to the greatest extent possible from other species within the community as more time becomes available. In contrast, the species with the lowest rate of advance, *Salix glauca*, achieves comparable segregation from other members of the community by not advancing its emergence timing. Both can be viewed as means to the same end: capitalizing on

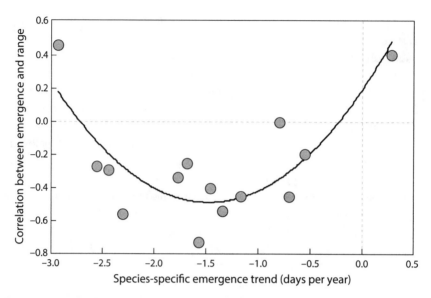

FIGURE 6.9. A test of whether the extent to which the correlation between species-specific emergence timing and the range of emergence dates within species is associated with species-specific trends in emergence dates. Here again, 12 years of data on species-specific emergence timing for the 14 species present in the plant community at the low-Arctic field site near Kanger-lussuaq, Greenland, are employed. Thus, each point represents metadata for a single species. Y-axis values are correlation coefficients quantifying the association between emergence dates and ranges of emergence dates for each species. Negative values are representative of the type of association depicted in figure 6.8c, but for species-specific emergence timing rather than community emergence timing, and indicate increased temporal segregation among individuals within species with earlier species-level emergence. X-axis values are trends in emergence timing for each species, and negative values thus indicate increasingly earlier species-level emergence. The parabolic nature of this association suggests that individuals of the species at the endpoints of the range of trends in emergence timing associate more closely in time with earlier emergence, perhaps thereby reducing temporal overlap with individuals of other species. The association is significant ($R^2 = 0.69$, $P < 0.05$).

an increase in the availability of time by reducing overlap in time with other species. In this context, both phenological advancement and phenological delay can be understood as adaptive, albeit opposing, strategies aimed at maximizing or optimizing use of time. In the following chapter, we will examine implications of the same types of dynamics for interactions among species in vertically structured communities characterized by consumer-resource associations. In chapter 8, we will examine patterns of phenological dispersion and segregation in tropical systems, where the availability of time is comparatively unconstrained by abiotic conditions.

CHAPTER SEVEN

Use of Time in the Phenology
of Vertical Species Interactions

In the preceding chapter, we explored the consequences for phenological dynamics at the species level of variation in the timing of expression of life history traits at the individual level. It was argued that if time is treated not only as an axis within the n-dimensional niche hypervolume but also as a resource in and of itself, then the alleviation of abiotic environmental constraints on the use of time should result in an increase in phenological dispersion among individuals within species. In other words, such alleviation should promote an increase in the segregation of individuals in time. But an increase in phenological dispersion among individuals within a species may also result in increasing phenological overlap among individuals of differing species. In a simple community comprising a single trophic level, it was argued, increases in phenological overlap among species in time may be disadvantageous. This would be the case in horizontal communities defined by lateral species interactions concerned with interference and competition for time or for other resources for which availability or access is constrained by time. Similarly, contrasting phenological dynamics among species in lateral communities, such as a delay by one species and an advance by another, may increase overlap in potentially competing species. Such opposing phenological trends have been documented in co-occurring species of amphibians in the southeastern United States (Todd et al. 2011).

In contrast, in more complex communities with multiple trophic levels phenological overlap in time is highly advantageous to mutualists and to consumers engaged in vertical species interactions concerned with exploitation of resources. Hence, minimizing competition for time by reducing phenological overlap should confer advantages for survival and reproduction in horizontally structured communities, but maximizing overlap with resource phenology should confer advantages for the survival and reproduction of consumers. When we consider resource species, however, the opposite is the case. For such species, minimizing phenological overlap with consumer species should be advantageous to their own survival and reproduction. The most challenging and complicated scenario is one in which a species is simultaneously compelled to maximize phenological over-lap with resource species, minimize phenological overlap with competitors, and minimize phenological overlap with consumers of itself (Both et al. 2009).

Before embarking upon an examination of the role of time in vertical species interactions, however, a brief reiteration of the concept of vertically structured communities is warranted. Vertically structured communities are those shaped primarily by interactions among organisms at different trophic levels. Hence, these comprise exploitation interactions typified by predator-prey interactions, pathogen host interactions, herbivore-plant interactions, and consumer-resource interactions in general (Post 2013). In such interactions, consumer success, in terms of growth, survival, and reproduction, depends upon synchronization of consumer phenology with resource phenology. In contrast, the success of resource species may depend upon minimizing synchronization of their phenology with that of species by which they are consumed. In mutualistic interactions, however, in which both species function as a resource for one another, the success of both species depends upon phenological overlap. Examples of the role of time in the phenology of all three types of players in vertical species interactions—resource species, consumer species, and mutualistic species—will be explored later.

THE USE OF TIME IN BITROPHIC-LEVEL SYSTEMS

Working from the simple, single-trophic-level community in figure 6.4, the conse-quences for both interference interactions and exploitation interactions of species-level shifts in phenology can be explored by adding a second trophic level to the system (figure 7.1). For simplicity, the bitrophic-level system in figure 7.1 comprises two species at each trophic level, but the dynamics explored here apply equally to any number of species at each level greater than two. Let us assume that each of the species in trophic level one in figure 7.1 represents a primary producer, and that each of the species in trophic level two represents a primary consumer. We may furthermore assume that the bars in figure 7.1 represent the range of dates of seasonal activity or presence of each of the species during the annual season of productivity in this system. Hence, the ends of the bars for the plant species might represent the 25th and 75th percentiles of the dates of onset and termination of growth or emergence, with the vertical lines within each bar representing the mean or median dates of such, and the whiskers representing an index of disper-sion of emergence dates such as standard errors of the means or the earliest and latest dates of emergence for each species. For the herbivore species (at trophic level two) in figure 7.1, the bars might represent comparable metrics related to the dates of presence at this particular site for each of the species.

Species within a two-level vertical community

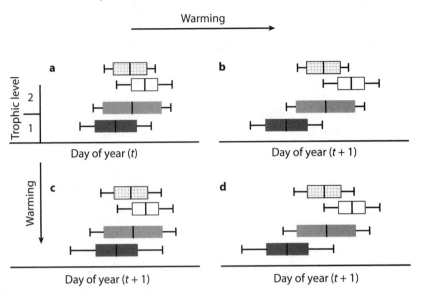

FIGURE 7.1. A hypothetical example of species-specific phenological dynamics following the alleviation of environmental constraints on the timing of expression of life history traits—in this case, warming—in a vertically structured community with two trophic levels. Trophic level one consists of two species of primary producers, and trophic level two consists of two species of primary consumers. Time within years progresses from left to right. Time progresses between years from baseline conditions in panel (a)—year *t*—to warmer conditions in panels (b), (c), and (d)—year *t* + 1 in all cases. Each box-and-whiskers plot represents the distribution of dates of activity (for example, growth or emergence for plants, presence at the site for herbivores) for each species. Vertical lines within boxes are median dates; left and right ends of boxes are the 25th and 75th percentiles, respectively, of the distribution of dates about the medians; and whiskers are the endpoints encompassing the first and last dates of activity. Hence, the phenological dispersion for each species is represented by the horizontal width of its box and by the distance between the whiskers at either end of its box. The progression from (a) to (b) represents differential, species-specific advancement of median dates of activity. Progression from (a) to (c) represents differential, species-specific changes in phenological dispersion without changes in median dates of activity. Progression from (a) to (d) represents a combination of both differential, species-specific advancement of median dates of activity and changing dispersion of dates about the medians.

Hence, in the baseline phenological state of this system (figure 7.1a), plant species one, depicted by the dark gray bar, tends to commence seasonal growth in advance of plant species two, represented by the light gray bar. In contrast, herbivore species two, represented by the stippled bar, tends to be present at the site earlier than herbivore species one, represented by the white bar. As well, herbivore species two appears to be well situated in time to exploit both plant species, and its presence is synchronized particularly closely with the seasonal growth of plant species two (figure 7.1a). By comparison, herbivore species one lags considerably behind the seasonal growth of plant species one, and appears better poised to exploit the phenology of plant species two (figure 7.1a). Let us now examine three possible outcomes of differential species-level responses within each trophic level to an environmental change that alleviates constraints on the use of time. As in the preceding chapter, although the example used here will relate to warming, alleviation of any operative abiotic constraint on phenology should precipitate comparable dynamics.

In the first example, moving from panel (a) to panel (b) in figure 7.1, warming has resulted in an advance in the timing of seasonal activity by both plant species, but a greater advance in species one than in species two. Neither species has undergone an alteration to the dispersion of dates of activity. At trophic level two, herbivore species two has advanced its timing of presence at the site in concert with the phenological advance of plant species two, while herbivore species one has not altered its phenology. Neither of the herbivores has experienced a change in the dispersion of dates of activity. Several consequences result from the differential shifts within each trophic level. There is reduced competition in time within each level as plant species one now becomes active considerably earlier than plant species two, and herbivore species two is present considerably earlier than herbivore species one. As well, plant species one is less vulnerable to exploitation by herbivore species two, and appears to have advanced its phenology to such an extent as to escape exploitation by herbivore species one entirely. Hence, the availability of relative ecological time represented by the phenology of plant species one has been reduced for both species of herbivores. In contrast, plant species two remains vulnerable to exploitation by herbivore species two because of the compensatory phenological shift by the latter, but likely experiences reduced exploitation by herbivore species one, which now lags somewhat behind the phenology of plant species two. Plant species one may be at the clearest advantage in this scenario by minimizing competition and reducing exploitation relative to the baseline scenario in panel (a).

In the next scenario, warming proceeds from panel (a) in figure 7.1 to panel (c). In this case, none of the species at either trophic level has advanced or delayed its phenology. Instead, both plant species have increased the dispersion of dates of activity about their respective means, with species one having done so to a greater extent than species two (see figure 7.1c). Hence, both species comprise some individuals characterized by earlier emergence and some individuals characterized by later emergence than was the case in the preceding time step. Among the herbivores, species two has undergone a similar change in response to warming, while species one remains unchanged. As a result, there is potential for greater competition in time at both trophic levels, but there is also potential for increased exploitation in time of plant species one by both herbivores as well as increased exploitation of plant species two by herbivore species one. Hence, herbivore species one in this case experiences an increase in the availability of relative ecological time as a result of changes in the phenology of both plant species. However, overall it is more difficult in this scenario to arrive at conclusions concerning advantages or disadvantages to particular species. Of greater relevance, in this case, are individual-level benefits of the changes to phenological dispersion. While later individuals may experience levels of interference and exploitation largely comparable to those in panel (a), the earliest individuals of both plant species and herbivore species two in panel (c) may experience the greatest benefits. Early individuals of both plant species may experience both reduced competition in time and reduced risk of exploitation by both herbivore species. Early individuals of herbivore species two are likely to experience greater success at tracking the phenologies of both plant species as well as reduced temporal overlap with herbivore species one.

The final warming scenario unfolds from panel (a) to panel (d) in figure 7.1. In this case, both plant species and herbivore species two have all advanced their mean timing of activity as well as the dispersion of dates about their respective means (figure 7.1c). Herbivore species one, in contrast, remains unchanged from the baseline scenario in panel (a). Here, again, plant species one has undergone a greater shift toward earlier emergence than has plant species two, as well as a greater increase in dispersion of emergence dates compared to species two. Hence, scenario (d) combines the dynamics depicted in panels (b) and (c). As a result, the consequences for each species, and individuals within them, comprise a combination of the consequences of the changes that apply to the scenarios represented in panels (b) and (c).

For instance, on average, plant species one should experience both reduced competition in time with plant species two, as well as reduced exploitation in time by herbivore species one and two. The benefits of such reductions apply most clearly to the earliest emerging individuals of plant species one. Plant species two, having remained closely tracked in time by herbivore species two, will, on average, experience exploitation by that herbivore comparable to that experienced in the baseline scenario, but reduced exploitation by herbivore species one. Among the herbivores, species two maintains its exploitation of plant species two, while lagging slightly more behind plant species one, and reduces, on average, competition in time with herbivore species one. Although herbivore species one now lags considerably more behind plant species one, the earliest individuals of herbivore species one maintain temporal overlap with the latest individuals of that plant species. Overlap by this herbivore with plant species two is maintained to some extent, although the latest individuals of herbivore species one now lag behind the emergence phenology of plant species two to such an extent that there is a complete mismatch in time. This represents a loss of relative ecological time. In the absence of any additional resource species advancing into this temporal "space," the disadvantages of this mismatch for late individuals of herbivore species one are likely to outweigh any advantages that might result from reduced competition in time with individuals of herbivore species two.

APPLICATIONS TO MUTUALISMS

Mutualisms represent a special case of vertical species interactions. In this case, the phenology of each of the species interacting across trophic boundaries is dependent upon the phenology of the species with which it interacts. In such interactions, the availability of relative ecological time is of paramount importance. As we saw in chapter 4, phenological mismatches resulting from differential rates or differential directions of shifts in the timing of biological activity between plants and pollinators may have negative consequences for persistence of such mutualisms and biodiversity of both plant and pollinator assemblages (Fabina et al. 2010; Bartomeus et al. 2011; Burkle and Alarcon 2011; Burkle et al. 2013; Bock et al. 2014; Forrest 2015). What do the scenarios depicted in figure 7.1 suggest when interpreted in the context of mutualisms? Let us revisit the consequences of phenological changes in response to warming, this time referring to the species at trophic level two as "pollinator one" in white and "pollinator two" in the stippled pattern. Let us furthermore consider the phenology of the two plant species as representative of the timing of flowering rather than emergence.

Starting with the baseline scenario presented in panel (a) of figure 7.1, the manner in which pollinator two tracks the flowering phenology of plant species two is highly suggestive of a specialized interaction. If we assume this is the case, then the persistence of both the plant and the pollinator seems, at the very least, unlikely to be threatened owing to phenological mismatch in response to warming, and at best, ensured in spite of warming. This is so because of the apparently perfectly compensatory tracking of shifts in the timing of flowering in plant species two by pollinator species two. In contrast, if we assume that pollinator species one is a generalist able to pollinate both plant species, then under the scenarios depicted in the changes from panel (a) to panel (b) and from panel (a) to panel (d), it appears likely that plant species one will lose the advantages to successful reproduction that accrue from interaction with this pollinator.

Furthermore, because pollinator species two is a specialist on plant species two, any remaining overlap in time with this pollinator in panels (b) and (d) presents no pollination services to plant species one. Hence, rather than solely benefiting from reduced competition in time with plant species two as it advances its phenology at a greater rate than its competitor in the scenarios described in the preceding section, in this context plant species one appears to have arrived at a disadvantage by advancing its timing of flowering at a rate that outpaces its pollinator. In further contrast to the scenarios presented in the preceding section, plant species two now appears to have benefited more as a consequence of maintaining temporal overlap with both pollinator species while simultaneously experiencing reduced competition in time with plant species one. As well, in all three scenarios depicted in panels (b), (c), and (d) of figure 7.1, the earliest individuals of both plant species are at the greatest disadvantage by having outpaced their pollinators in time. For them, relative ecological time has been lost. Last, the outcomes of the phenological shifts resulting in the scenarios in panels (b) and (d) appear most likely to place pollinator species one at greatest risk of local extinction owing to the loss of floral resources resulting from phenological stasis by this pollinator. Indeed, as we saw in chapter 4, evidence for adverse consequences of phenological mismatch between plants and pollinators abounds (Aizen and Rovere 2010; McKinney and Goodell 2011; McKinney et al. 2012; Høye et al. 2013).

THE USE OF TIME IN TRITROPHIC-LEVEL SYSTEMS

Next, let us consider a slightly more complex system by adding a third trophic level, but in the simplest manner possible, with a single species of secondary consumer (figure 7.2). Assume that the bars in figure 7.2a represent, for trophic levels one and two, respectively, the seasonal timing of growth or emergence in two species of plants and the presence of two species of herbivores at the same site. Hence, the phenological dynamics described previously for this system and represented graphically in figure 7.1 apply here in identical fashion. Additionally, however, we now include a bar, at trophic level three, representing the seasonal timing of the presence of a predator at the site.

In the baseline scenario depicted in panel (a) of figure 7.2, the timing of the onset and termination of the presence of the predator at the site is bracketed entirely within the timing of the presence of herbivore species two at the site, and overlaps to a considerable extent the timing of the presence of herbivore species one. Let us assume that this is a generalist predator, able to track in time and depredate both species of herbivores. In the first warming scenario,

Species within a three-level vertical community

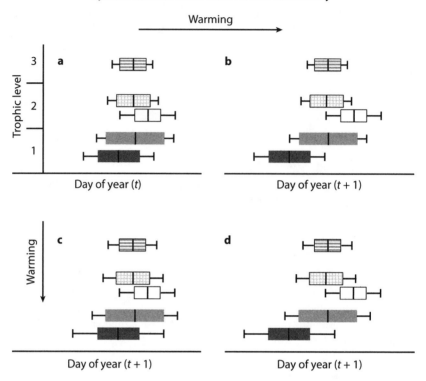

FIGURE 7.2. Another hypothetical example of species-specific phenological dynamics following the alleviation of environmental constraints on the timing of expression of life history traits—in this case, warming—in a vertically structured community, as in figure 7.1, but in this case representing dynamics in a tritrophic-level community. Here, the system includes two species of primary producers (dark and light gray bars; bottom set in each panel), two species of primary consumers (herbivores, white and hatched bars; middle set in each panel), and a single species of secondary consumer (predator, gray striped bar; top in each panel). Time progresses between years from baseline conditions in panel (a)—year t—to warmer conditions in panels (b), (c), and (d)—year $t + 1$ in all cases. Each box-and-whiskers plot represents the distribution of dates of activity (for example, growth or emergence for plants, presence at the site for herbivores and the predator) for each species. Vertical lines within boxes are median dates; left and right ends of boxes are the 25th and 75th percentiles, respectively, of the distribution of dates about the medians; and whiskers are the endpoints encompassing the first and last dates of activity. Hence, the phenological dispersion among individuals within each species at each trophic level is represented by the horizontal width of its box and by the distance between the whiskers at either end of its box. As in figure 7.1, the progression from panel (a) to panel (b) represents differential, species-specific advancement of median dates of activity. Progression from (a) to (c) represents differential, species-specific changes in phenological dispersion without changes in median dates of activity. Progression from (a) to (d) represents a combination of both differential, species-specific advancement of median dates of activity and changing dispersion of dates about the medians.

that depicted as the progression from panel (a) to panel (b) in figure 7.2, both plant species and herbivore species two advance the timing of emergence and presence, respectively, while herbivore species one displays phenological stasis, exactly as in panel (b) of figure 7.1, earlier. Additionally, the predator advances its timing of presence at the site, appearing to track perfectly the phenological shift exhibited by herbivore species two. As a consequence, herbivore species one now experiences reduced overlap in time with its predator. Consequently, whereas previously in the bitrophic-level system, phenological stasis by herbivore species two may have been largely disadvantageous because of the resulting reduction in temporal overlap with both plant species (see figure 7.1b), in this scenario it is likely that herbivore species one will derive some benefit from reduced exposure in time to predation. Hence, in this case, we observe the use of time as a refuge to the extent that this part of the season is now largely predator-free.

Revisiting the next scenario, from panel (a) to panel (c) in figure 7.2, recall that in this case none of the species in this system advanced their mean or median timing of presence at the site. Rather, with the exception of herbivore species one, all species underwent an increase in the dispersion of dates of presence about the mean or median, with some individuals commencing growth or arriving at the site earlier, and some later, in comparison to the baseline scenario in panel (a). Additionally, in this case, the predator has undergone an increase in the dispersion of its dates of presence at the site, once again appearing to maintain temporal overlap with herbivore species two. In the same scenario as applied to the bitrophic-level system, this increase in phenological dispersion was interpreted as presenting some benefit to herbivore species one because of increased overlap in time with plant species one. It was also interpreted as a disadvantage for herbivore species one owing to slightly increased temporal overlap with its competitor, herbivore species two, later in the season. In the tritrophic-level system, these interpretations apply as well, but now the potentially disadvantageous increase in competition in time with herbivore species two is compounded by increased exposure to predation risk for herbivore species one late in the season.

Last, in the warming scenario from panel (a) to panel (d) of figure 7.2, we see advances in the timing of growth or presence at the site by both plant species and by herbivore species two, as well as increases in the dispersion of dates about the means or medians for all three species. The phenology of herbivore species two remains unchanged from the baseline conditions. And here, the predator has undergone both an advance in the timing of its presence at the site as well as an increase in the dispersion of its own dates of presence at the site (figure 7.2c). Previously, in the bitrophic-level system, these changes were interpreted as potentially presenting herbivore species one with advantages related to somewhat reduced competition in time with herbivore species two and sustained overlap in time with plant species two. Now, however, there is a potential disadvantage to the earliest individuals of herbivore species one that derives from continued temporal exposure to predation risk. Such dynamics are highly suggestive of the potential for development of a phenomenon we might term a "phenological cascade."

PHENOLOGICAL CASCADES

Cascades in ecology are represented by changes at one trophic level induced by changes in adjacent trophic levels (Strong 1992). The most widely documented form of ecological cascade is the trophic cascade, in which changes in the abundance or productivity of primary producers result from changes in abundance or behavior of primary consumers, or in which changes in abundance at higher trophic levels result from changes in primary productivity (Carpenter et al. 1985; Power et al. 1985; Carpenter and Kitchell 1988; Power 1992; Pace et al. 1999; Carpenter et al. 2001, 2011). Changes in the abundance or behavior of primary consumers may also, in turn, result from depredation by secondary consumers or even from their mere presence in the system in a special type of cascade known as the behaviorally mediated trophic cascade (Schmitz 1998; Schmitz et al. 2000). In such tritrophic-level systems, trophic cascades can elicit far-reaching effects on ecosystem structure and function (Post et al. 1999b; Schmitz et al. 2003; Schmitz 2004, 2008, 2010). If, as we have seen, changes in phenological dynamics have the potential to alter interactions among species, then what might be the consequences of changes in phenology of species at one trophic level for the phenology of species at adjacent and interacting trophic levels? The answer to this question depends, of course, on the trophic level from which the phenological change initiates. Generally, however, such phenological cascades may take two forms. In a direct phenological cascade, a shift in the timing of life history activity at one trophic level elicits a shift in timing of life history activity at an adjacent, interacting, trophic level. In an indirect phenological cascade, a shift in timing of life history activity at one trophic level elicits a change in abundance or some other demographic trait such as productivity at an adjacent, interacting, trophic level. Next, we will examine phenological cascades emanating from alterations to the phenology of primary producers in bitrophic- and tritrophic-level systems, the phenology of primary consumers at the top of or embedded in the middle of bitrophic- and tritrophic-level systems, and the phenology of primary consumers atop tritrophic-level systems.

EMPIRICAL ASSESSMENT OF PHENOLOGICAL
DYNAMICS IN BITROPHIC-LEVEL SYSTEMS

In the preceding chapter, we examined the consequences of species-specific phenological shifts for overlap in time both within and among species that potentially engage in competition at a single level within a community. Empirical data from multiannual observations of plant phenology at the long-term study site near Kangerlussuaq, Greenland, were used to illustrate consequences of variation in species-specific trends in emergence phenology for overlap in time among species in the local assemblage. The Kangerlussuaq system is functionally bitrophic because it includes resident species of large vertebrate herbivores, caribou and muskoxen, but lacks large predators of these herbivores. Arctic foxes, white-tailed sea eagles, and ravens at the site may depredate caribou calves on occasion, but they are incapable of killing adult caribou or muskoxen, and their influences on population dynamics of large herbivores at the site are likely negligible. Hence, for the purpose of examining phenological dynamics in a vertically structured, bitrophic-level system that can be linked to phenological dynamics within a horizontally structured single level within this system, consider caribou and muskoxen at the Kangerlussuaq site.

Numbers of caribou and muskoxen observed within a fixed perimeter of the Kangerlussuaq study site have been recorded on a daily or near-daily basis immediately preceding and throughout the spring and early summer field seasons beginning in 1993 and annually since 2002 (Post 1995; Bøving and Post 1997; Post et al. 2003; Post and Forchhammer 2008; Post and Pedersen 2008; Post et al. 2008b; Kerby and Post 2013a, 2013b; Post 2013). These data provide a record of the timing of peak presence of both species in the calving area that is the core of the study site (John 2016). Muskoxen are resident year-round in the vicinity of the study site, but they, like their conspecifics at the Zackenberg study site in Northeast Greenland (Forchhammer et al. 2005, 2008a, 2008b), engage in a form of rotational migration into and out of the core study area at Kangerlussuaq. By contrast, caribou in the area migrate from winter ranges west of the Kangerlussuaq study site into the main calving area at the core of the study site in spring, and then migrate out of it again around mid- to late summer (Thing 1984; Meldgaard 1986; Bøving and Post 1997). Hence, fluctuations in the numbers of muskoxen observed at the site relate to their use of dispersed foraging areas in and around the main site, while fluctuations in the numbers of caribou at the site reflect seasonal migrations into and out of the area preceding and following parturition (John 2016).

By converting the cumulative total numbers of caribou and muskoxen observed each day to the cumulative proportion of the total observed each day, we can quantify the phenology of arrival timing, or of the peak presence, of each species for each year of observation (John 2016). To do this, we apply equation (6.3) to the daily proportion of the total numbers of caribou or muskoxen observed each year to derive year-specific parameter estimates of the model. These can then be solved to obtain estimates of the dates on which any proportion or percentage of the cumulative total numbers of each species were observed each year. To quantify the midpoint of the season of arrival or migration into the area, we solve equation (6.3) for 0.50, or the date on which 50 percent of the total count of each species was obtained in each year. Additionally, we may compare the dates on which the peak numbers of each species were observed in each year. To derive a range of annual arrival or migration timing, we can solve equation (6.3) for 0.25 and 0.75, or the dates on which 25 percent and 75 percent, respectively, of the total count of each species was obtained in each year. The difference between these latter two dates then provides a range of arrival or migration timing among individuals of each species centered on the species-specific annual date of 50 percent arrival. This range, an index of phenological dispersion of migration or arrival dates, can then be used to quantify overlap in the timing of presence at the site between caribou and muskoxen in a manner comparable to that in which phenological overlap among plant species in their timing of emergence was achieved in the preceding chapter.

Applying this methodology to the annual daily counts of caribou and muskoxen at the Kangerlussuaq study site, we notice, first, considerable variation in the timing of presence of both species at the site (figure 7.3). For both species, the annual dates of 50 percent presence at the site are represented in figure 7.3 with black dots, while the annual dates of peak presence (when the maximum numbers of individuals of each species were recorded) are represented with white dots. For caribou, the midpoint of arrival on the calving grounds, the annual date on which 50 percent of the total cumulative number of individuals is observed, has occurred as early as day 146 (May 25, 2012) and as late as day 170 (June 19, 2011) (John 2016). Similarly, peak presence of caribou has occurred as early as day 147 (May 27, 2014) and as late as day 173 (June 22, 2011) (John 2016). For muskoxen, the annual date of 50 percent presence on the calving grounds has varied from as early as day 139 (May 18, 2012) to as late as day 166 (June 15, 2006). The annual date of peak presence of muskoxen at the site has occurred as early as day 142 (May 22, 2010) and as late as day 170 (June 19, 2006).

While dates of 50 percent presence and peak presence are not significantly correlated in caribou ($r = 0.41$, $P = 0.15$), they are in muskoxen ($r = 0.76$, $P = 0.002$). Moreover, muskoxen tend to achieve 50 percent presence and peak annual presence significantly earlier than caribou, on average by 3.5 days and 6.1 days, respectively (both mean pairwise difference tests are significant at $P = 0.02$). Despite the variation in arrival timing by caribou, neither the date of peak presence nor the

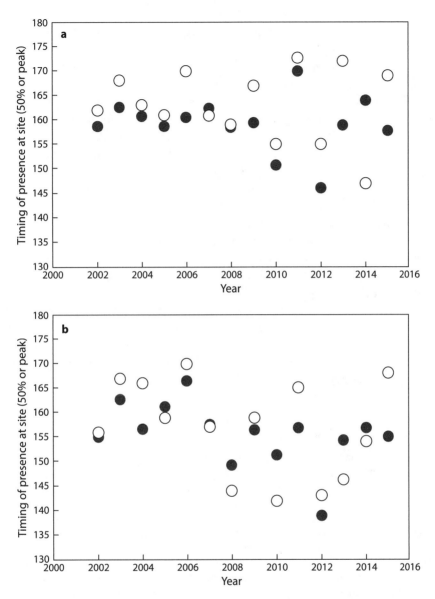

FIGURE 7.3. Interannual variation in the timing of presence (an index of migratory arrival) of caribou (a) and muskoxen (b) at the study site near Kangerlussuaq, Greenland, from 2002 through 2015. Beginning at or near the annual onset of the plant-growing season, numbers of individuals of both species of herbivore occurring within the study site are counted on a daily or near daily basis along fixed census routes. These counts are used to estimate the annual date at which 50 percent presence (black dots) and peak presence (white dots) of each species is achieved. The date of 50 percent presence represents the midpoint of the seasonal influx of individuals into the site, while the date of peak presence is defined as the date on which the annual maximum number of individuals was observed. Units along the y-axis are the day of year.

date of 50 percent presence display a significant trend, though both correlations with "year" are weakly negative ($r_{peak} = -0.12$ and $r_{50\%} = -0.15$, respectively) (John 2016). Stronger negative trends exist for muskoxen, but neither of these is significant either ($r_{peak} = -0.31$ and $r_{50\%} = -042$, respectively; see figure 7.3).

What are the consequences of variation in the timing of migration phenology by both caribou and muskoxen into the core calving area for temporal overlap between the two species at the site? Plotting data for the annual dates of 25 percent, 50 percent, and 75 percent presence at the site by both species permits visual examination of temporal overlap of arrival phenology in caribou and muskoxen, as well as of the magnitude of phenological dispersion in arrival timing within each species and the extent to which this varies among years (figure 7.4a). Here, the dispersion in arrival timing within each year for each species is represented as the range of days between the dates of 25 percent presence and 75 percent presence, illustrated in figure 7.2a as bars below and above, respectively, the annual dates of 50 percent presence. This plot reveals almost complete overlap in the timing of 50 percent presence at the site by both species in some years, notably 2003 and 2010, and nearly complete temporal segregation of the species in other years, such as 2008 and 2011 (figure 7.4a). Moreover, remarkably narrow dispersion, characterized by tight bounds of the dates of 25 percent and 75 percent presence, in both species in years such as 2002 and 2005 stands in contrast to marked dispersion in other years, notably 2011 and 2014.

FIGURE 7.4. (a) Interannual variation in timing of presence of caribou (black dots) and muskoxen (gray dots), as well as in phenological dispersion among individuals within each species, at the study site near Kangerlussuaq, Greenland, from 2002 through 2015. Here, timing of presence is given as the annual date of 50 percent presence at the study site. The bars below and above, quantifying the dispersion of dates of presence about the date of 50 percent presence, represent the annual dates of 25 percent and 75 percent presence, respectively. Hence, in addition to the marked variation among years in the timing of 50 percent presence in each species, there is considerable variation from year to year in the dispersion of dates of arrival at the study site about the annual date of 50 percent presence. The degree of phenological dispersion of arrival dates within each species is an index of intraspecific overlap in time, and, therefore, of potential intraspecific competition in time. (b) The degree of dispersion within each species in each year also, however, influences the index of overlap in time between the two species. This index of interspecific phenological overlap is calculated as the minimum value of the annual date of 75 percent presence of the two species minus the maximum value of the annual date of 25 percent presence of the two species in the same year. Interspecific phenological overlap between caribou and muskoxen at the site has increased over the course of the study ($r = 0.68$, $P < 0.01$). Y-axis values in panel (a) are the day of year, and those in panel (b) are the number of days of phenological overlap.

Quantification of the degree of interspecific overlap between the two species may be achieved in a comparable manner to that applied in the preceding chapter for overlap in emergence timing of plants at the site. Here, we take the difference between the minimum of the respective dates of 75 percent presence for each species in a given year and the maximum of their respective dates of 25 percent

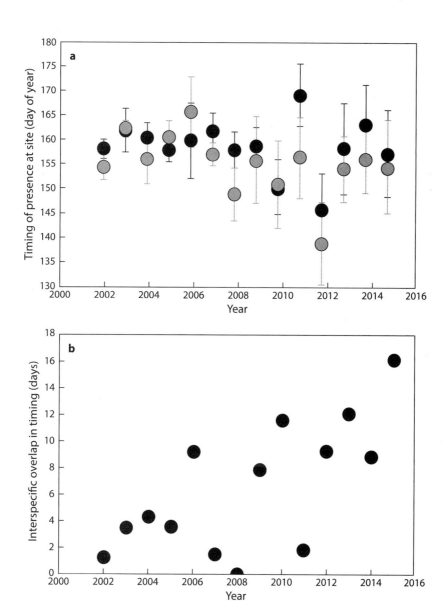

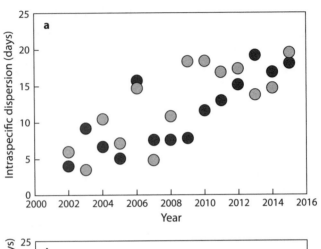

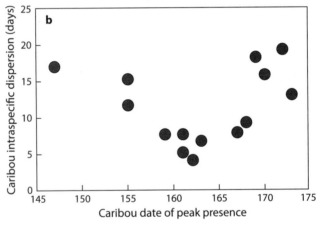

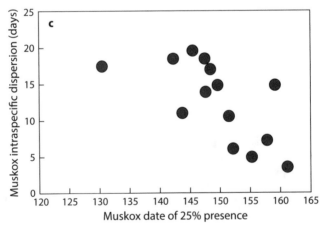

presence in the same year. This provides a quantitative index of phenological overlap between the two species at the site that can be examined for interannual variation. This index displays a significant positive trend over the period of observation to date ($r = 0.68$, $P < 0.01$), indicating increasing temporal overlap of caribou and muskoxen at the site (figure 7.4b). Presumably, this relates to variation in intraspecific dispersion of dates of presence at the site within each species because, as noted earlier, neither species has undergone a significant trend in dates of 50 percent presence or peak presence. Both species do, however, display positive trends in intraspecific dispersion of dates of presence at the site (figure 7.5a) ($r_{caribou} = 0.82$, $P < 0.001$; $r_{muskoxen} = 0.77$, $P = 0.001$).

These patterns indicate increasing phenological segregation among individuals of each species in the timing of arrival at the site, despite the lack of trends in timing of peak presence or 50 percent presence for either species. To explain this variability, we can apply an analysis testing for the significance of several potential predictor variables. For caribou, variation in the degree of phenological dispersion in dates of presence at the site is related only to variation in dates of peak presence. Interestingly, this relationship is best fit by either a quadratic or a cubic spline function ($R^2 = 0.57$ and $P = 0.01$ in both cases). The nature of this relationship suggests that the greatest phenological dispersion in dates of presence at the site by caribou occurs in years of very early or very late peak presence (figure 7.5b). In contrast, for muskoxen, variation in the degree of dispersion in dates

---

FIGURE 7.5. (a) Interannual variation in the degree of intraspecific phenological dispersion in caribou (black dots) and muskoxen (gray dots), at the study site near Kangerlussuaq, Greenland, from 2002 through 2015. Here, as in figure 7.4, intraspecific phenological dispersion represents variation in dates of presence at the study site about the annual date of 50 percent presence for each species. Trends are significant for each species (see main text), indicating that the range of dates over which each species arrives at the site in spring has increased over the course of the study. For caribou, this trend is significant despite the lack of significant trends in dates of 25 percent, 50 percent, or 75 percent presence at the site (figures 7.3a and 7.4a). (b) The aspect of caribou arrival phenology that displays a significant association with intraspecific phenological dispersion is the annual date of peak presence at the site (see main text), but this association is strongly nonlinear, indicating that dispersion is greatest in years of earliest and latest arrival. For muskoxen, the trend toward increasing intraspecific phenological dispersion is significant despite the lack of significant trends in dates of 50 percent presence, 75 percent presence, and peak presence (figures 7.3b and 7.4a). (c) Rather, increasing phenological dispersion in muskoxen relates strongly to the trend toward earlier presence driven by annual dates of 25 percent presence (see main text). Hence, earlier onset of presence of muskoxen at the study site promotes reduced overlap in time among members of this species within each season. X-axis values in (b) and (c) are the day of year, while y-axis values in (b) and (c) are number of days of overlap within each year.

of presence at the site relates only to variation in the date of 25 percent presence (figure 7.5c; $r = -0.65$, $P = 0.01$). In other words, early occurrence of muskoxen at the site is characterized by greater variation in dates of presence, while the opposite applies in years of late arrival.

But what drives interspecific overlap between caribou and muskoxen in their timing of presence at the site? Recall that phenological overlap between the two

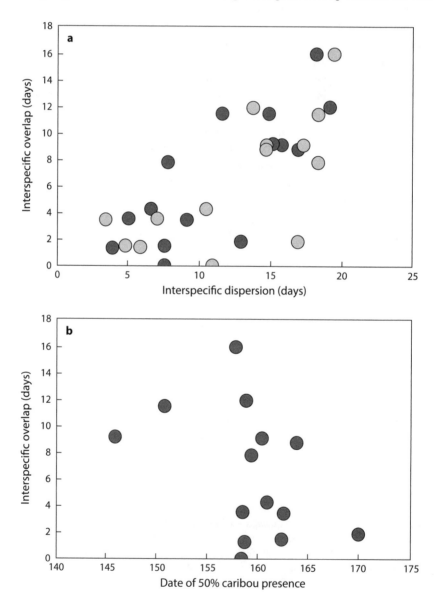

species has increased at the site over the course of the study to date (figure 7.4b). As predicted, the intraspecific dispersion indices of both caribou and muskoxen correlate positively with the degree of temporal overlap between the two species at the site (figure 7.6a) ($r_{caribou} = 0.78$, $P < 0.01$; $r_{muskoxen} = 0.71$, $P < 0.01$). We should be cautious, however, in correlating multiple phenological time series when all of them display trends, because of the potential for spurious correlations driven by some other factor that itself is related to time (Forchhammer et al. 1998; Post and Stenseth 1998, 1999). A more robust analysis would include multiple candidate predictor variables, and a term for "year" itself, to account for trends in each of the time series (Forchhammer et al. 2002; Forchhammer and Post 2004). With time series of sufficient length, a proper analysis could be undertaken including the application of autoregressive time series models that account for temporal dependence among successive years of data (Forchhammer et al. 1998; Post et al. 2001a), but the length of the time series is, in this case, potentially insufficient for such an approach. Among the candidate predictor variables for consideration in an analysis of variation in interspecific overlap in timing of presence of caribou and muskoxen at the study site, indices of intraspecific temporal dispersion for each species, as well as their respective dates of 50 percent presence and peak presence seem appropriate. We might also consider including dates of 25 percent and 75 percent presence for each species, but the index of interspecific overlap itself, as well as the species-specific indices of temporal dispersion, are calculated using these dates, and so exclusion of them as candidate predictor variables is warranted.

Such an analysis reveals that, after accounting for trends in the data, only two variables emerge as significant predictors of interspecific temporal overlap: intraspecific dispersion in caribou (partial $P = 0.001$) and the annual date of 50 percent

---

FIGURE 7.6. Factors associated with interannual variation in the degree of interspecific phenological overlap between caribou and muskoxen in the annual timing of their presence (arrival) at the study site near Kangerlussuaq, Greenland, from 2002 through 2015. In panel (a), the degree of interspecific overlap between the two species in a given year is plotted against the degree of intraspecific phenological dispersion in that year for caribou (black dots) and muskoxen (gray dots). The degree of interspecific phenological overlap correlates with the degree of intraspecific phenological dispersion in both caribou ($r = 0.78$, $P < 0.05$) and muskoxen ($r = 0.71$, $P < 0.05$). However, these simple correlations do not account for trends in both independent and dependent variables. In panel (b), annual degree of interspecific phenological overlap is plotted against annual date of 50 percent presence at the study site by caribou. This correlation is not significant ($r = -0.42$, $P > 0.05$), but the partial correlation between these two variables is significant after detrending the time series and accounting for the partial correlation with the degree of intraspecific phenological dispersion in caribou (see main text). Hence, earlier presence of caribou at the site contributes to greater overlap in time between caribou and muskoxen during each season of observation.

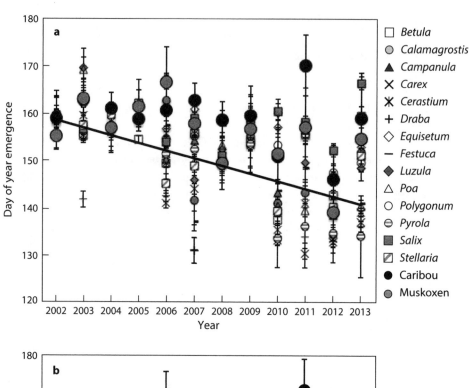

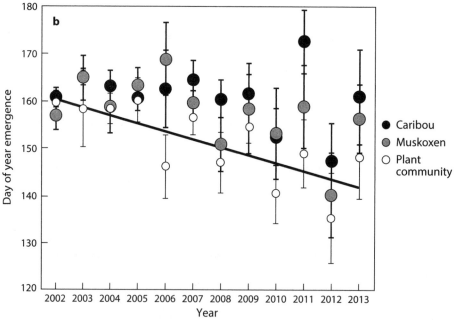

presence in caribou (partial $P = 0.05$). The latter partial correlation, depicted in figure 7.6b as a plot of interspecific overlap versus the raw data for dates of 50 percent caribou presence rather than as a residual plot, further indicates that early presence of caribou at the site contributes to greater temporal overlap between muskoxen and caribou.

But these analyses have thus far ignored any potential associations between herbivore phenology and that of their resource base, plants. Because of this, we have considered only consequences of changes in absolute timing—that is, changes in arrival timing in and of itself—for intraspecific phenological dispersion and temporal overlap between caribou and muskoxen. Let us now turn our attention to patterns of variation in the timing of presence at the study site by these two large herbivores, and of overlap between them, in relation to patterns of variation in the timing of emergence of forage plant species and temporal overlap among species at the primary producer level in this system. This will place the emphasis on an examination of consequences for herbivore phenology of changes in *relative ecological time* represented by shifts in the timing of resource availability, suggestive of a phenological cascade.

Returning our attention to species-specific and community-level plant phenological dynamics at the Kangerlussuaq study site illustrated earlier in figure 6.5, the arrival phenology of caribou and muskoxen has, in figure 7.7a, been superimposed on the same plant phenology data. Note that, although data on the arrival phenology of caribou and muskoxen extend through 2015, we have plotted data through the year 2013 in this case because that is the last year of published data for plant phenology at the site. As noted in the previous chapter, a significant trend in community-wide emergence timing is evident, illustrated again in figure 7.7a as the black trend line. This trend represents progressively earlier occurrence of the midpoint of the season of plant emergence, the annual date on which 50 percent of species have emerged (Post and Forchhammer 2008; Kerby and Post 2013b). In contrast, as noted earlier, neither the midpoint of the timing of arrival by caribou, nor the midpoint of the timing of arrival by muskoxen into the site, the

---

FIGURE 7.7. Interannual variation in the timing of presence (arrival) of caribou (black dots) and muskoxen (gray dots) at the study site near Kangerlussuaq, Greenland, in relation to resource phenology. In panel (a), mean (±1 SE) annual dates of emergence of all 14 species of plants occurring on long-term phenological monitoring plots are overlain by annual dates of 50 percent (± 25 percent) presence of caribou (black dots) and muskoxen (gray dots) at the site. The black trend line is fit to estimates of the annual date of 50 percent emergence of plant species calculated across the entire community. In panel (b), plant species-specific mean annual dates of emergence have been replaced with community-level annual dates of 50 percent (± 25 percent) emergence.

dates on which each of these species has reached 50 percent presence at the site, has advanced significantly over the course of the study.

Despite the congested appearance of figure 7.7a, owing to the inclusion of data on species-specific plant emergence timing, it is possible to discern a pattern of the outpacing of herbivore arrival timing by advancing plant emergence phenology at the community level. Nonetheless, when we examine plant species-specific phenological dynamics, the timing of caribou arrival at the site correlates most closely, and significantly, with the timing of emergence of two forbs, *Campanula* sp. ($r = 0.83$, $P < 0.001$) and *Stellaria* sp. ($r = 0.77$, $P < 0.01$), and the deciduous shrub, *Betula nana* ($r = 0.71$, $P = 0.01$). Timing of arrival at the site by muskoxen, in contrast, correlates most closely and significantly with the timing of emergence of two forbs, *Equisetum* sp. ($r = 0.83$, $P < 0.001$) and *Campanula* sp. ($r = 0.75$, $P = 0.005$), and two graminoids, *Carex* sp. ($r = 0.76$, $P = 0.004$) and *Poa* sp. ($r = 0.73$, $P < 0.01$).

This plot can be tidied up a bit by summarizing the plant phenology data using metrics of community-wide emergence timing. In panel (b) of figure 7.7, the midpoint of community-wide emergence, the date on which 50 percent of species have emerged, is represented by the white dots to which the trend line is fit. The lower and upper endpoints of the bars centered on the annual dates of 50 percent emergence represent annual dates of 25 percent and 75 percent community-wide emergence, respectively. Similarly, the gray and black dots in figure 7.7b represent the annual dates of 50 percent arrival at the site by muskoxen and caribou, respectively, while the lower and upper endpoints of the bars centered on those dots represent the annual dates of 25 percent and 75 percent arrival for each species. Hence, the phenology of the primary producer level of this system can now be compared directly to the phenology of the herbivore level.

From an initial visual inspection of these time series, it is immediately apparent that the extent of overlap between herbivore phenology and primary producer phenology varies considerably among years. Furthermore, the species of herbivore that experiences the most complete or the most incomplete overlap with primary producer phenology also varies from year to year. For instance, consider the contrast among years 2002, 2004, and 2005. In the first of these years, primary producer phenology falls nearly entirely between muskoxen arrival and caribou arrival at the site. In year 2004, the timing of muskoxen arrival at the site coincides nearly perfectly with the midpoint of the community-level season of emergence, while caribou arrival lags behind both. In contrast, in year 2005, caribou arrival phenology overlaps plant emergence almost perfectly, while the arrival timing of muskoxen lags behind both. In the following year, 2006, the midpoint of the community-level emergence season, as well as the 25 percent and 75 percent endpoints, occur earlier than, and are entirely disassociated in time from, the arrival of both herbivore species at the site. The year 2006 also happened to be one of the three earliest years of plant emergence at the site (figure 7.7a).

This suggests that, at least in some years, one or both species of large herbivores in this system may not be capable of advancing their arrival phenology to the extent of entirely matching increasing availability of relative ecological time as resource phenology advances. Before examining the factors that are associated with variation in the extent to which each of these two herbivore species overlaps in time with the phenology of its resource base, let us take a moment to explore the timing, intraspecific phenological dispersion, and interspecific phenological overlap of caribou and muskoxen in relation to plant phenology. This pursuit will, it turns out, provide an empirical example illustrative of a direct phenological cascade.

AN EMPIRICAL EXAMPLE OF A PHENOLOGICAL CASCADE

Among the potential candidate phenological predictors of the timing of herbivore arrival at the site, the onset of the community-wide season of emergence is the strongest. For both caribou and muskoxen, the annual date of 25 percent arrival at the core calving area is positively associated with the annual date of 25 percent community-level plant species emergence (figure 7.8a), despite the absence of significant trends for either species in the annual timing of the midpoint or peak of arrival. The magnitudes of these correlations are comparable, although slightly stronger for caribou ($r_{caribou} = 0.76$, $P = 0.004$; $r_{muskoxen} = 0.72$, $P = 0.006$), and suggest that earlier onset of the plant growing season is matched by earlier arrival of both species of herbivores at the site. Importantly, however, in agreement with the pattern inferred from figure 7.7, the timing of arrival of both herbivores at the site tends to lag the timing of onset of the plant community-level season of emergence, most notably in years of early plant emergence (see also John 2016). This is evident in figure 7.6a as the generally positive deviation from the line of parity by the dates of 25 percent arrival by caribou and muskoxen. In years of later plant emergence, the timing of arrival by muskoxen tends to cluster about the parity line, and in some years muskoxen arrive in advance of the date of 25 percent plant emergence, while caribou continue to lag behind this date (figure 7.8a).

Next, consider the implications for intraspecific phenological dispersion at the herbivore trophic level of advancing plant phenology or earlier onset of resource availability. Recall from the discussion earlier in this chapter that intraspecific phenological dispersion is an index of the magnitude of overlap in time among conspecifics, and therefore also of potential competition in time among conspecifics. In both caribou and muskoxen, this index was related to the timing of presence of each species at the site, though in different ways (see figures 7.5b,c). If the

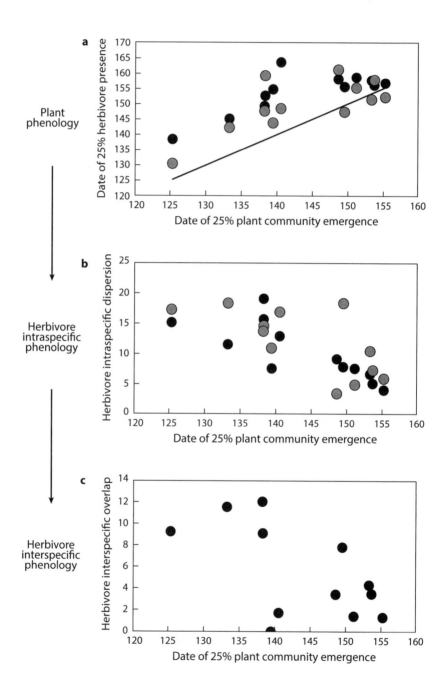

Plant
phenology

Herbivore
intraspecific
phenology

Herbivore
interspecific
phenology

timing of arrival of both herbivores is related to the timing of onset of the plant growing season, we should expect intraspecific phenological dispersion within each species to be associated with plant phenology as well. Indeed, significant correlations exist between the date of 25 percent plant community emergence and phenological dispersion within both caribou and muskoxen (figure 7.8b). In both cases, the correlation is negative, though slightly more strongly so in caribou than in muskoxen ($r_{caribou} = -0.78$, $P = 0.003$; $r_{muskoxen} = -0.69$, $P = 0.01$). Hence, for both species, earlier onset of the community-level season of plant emergence is associated with earlier arrival at the site and, in turn, with greater variation within each species in the timing of arrival. This pattern suggests a tendency in both species to capitalize upon increasing availability of relative ecological time by reducing temporal overlap with conspecifics. In contrast, years of reduced availability of relative ecological time, characterized by a later onset of the plant community-level season of emergence, are typified by greater temporal overlap among conspecifics of both species (that is, reduced phenological dispersion; figure 7.8b).

Last, consider the consequences of changes in the availability of relative ecological time for interspecific overlap in time between both species of herbivore. As noted earlier, temporal overlap between caribou and muskoxen at the site tends to increase as intraspecific phenological dispersion increases (figure 7.6a). In other words, as individuals of each species segregate in time from members

FIGURE 7.8. An example of a direct phenological cascade, in which phenological dynamics at one trophic level elicit changes in phenological patterns at adjacent trophic levels. Here, the cascade involves the bitrophic-level system at the study site near Kangerlussuaq, Greenland, comprising plants and herbivores. The cascade begins in panel (a) with an association between the timing of onset of the presence of the two species of herbivore at the site (caribou, black dots; muskoxen, gray dots) and the timing of onset of the plant-growing season at the site. The timing of onset is quantified as the annual date of 25 percent presence for caribou and muskoxen and the annual date on which 25 percent of the plant species in the plant community have emerged. In panel (a), the black line represents parity—that is, synchrony between the timing of onset of herbivore presence and onset of plant growth. Hence, the timing of occurrence of caribou at the site generally lags behind the onset of plant growth, while the presence of muskoxen at the site tends to be synchronous with the onset of plant growth in years characterized by a later start to the growing season. Panel (b) illustrates that years of early onset of plant growth tend to be characterized by greater intraspecific phenological dispersion in both caribou (black dots) and muskoxen (gray dots), while this dispersion diminishes in years of later onset of plant growth. The correlations in panel (b) are significant for both caribou and muskoxen (see main text). As a consequence of increased intraspecific phenological dispersion, as shown in panel (c), years characterized by early onset of the plant-growing season are also characterized by greater overlap in time between caribou and muskoxen at the study site, as indexed by interspecific phenological overlap ($r = -0.59$, $P < 0.05$).

of their own species, a strategy that presumably reduces the potential for intra-specific competition in time, they tend to increase the magnitude of temporal overlap with members of the other species, presumably increasing the potential for interspecific competition in time. The analysis presented earlier revealed that the extent of interspecific overlap in time tends to relate to the timing of caribou arrival at the site. Earlier caribou arrival coincides with increased temporal overlap between caribou and muskoxen. When variation in the phenology of the resource base is included in this picture, we observe a significant association between the timing of the onset of the plant community-level emergence season and the degree of phenological overlap between caribou and muskoxen (figure 7.8c; $r = -0.59$, $P = 0.04$). Hence, earlier onset of the plant growing season, considered in this context as representative of an increase in the availability of relative ecological time, precipitates a phenological cascade characterized by the following sequence of consequences, all of which are related to timing. First, both species of herbivores occur earlier at the site (figure 7.8a). As a consequence, each of them displays an increase in intraspecific dispersion in the timing of arrival at the site (figure 7.8b). And as a result of this, the extent of overlap in time between the two species of herbivores increases in turn (figure 7.8c). But what are the consequences, if any, of variation in plant phenology for trophic overlap in time in this vertically structured community between each species of herbivore and the plant community?

The data in figure 7.7, and the species-specific correlations deriving from the data in figure 7.7a in particular, have already been used to infer variation in the extent to which herbivore presence at the site correlates with the timing of plant emergence. Next, however, an analysis of phenological overlap between caribou and the plant level of the community, and similarly between muskoxen and the plant level of the community, is necessary to address this question. As was done in the preceding analysis, phenological overlap between each species of herbivore and the plant community can be quantified using the ranges of dates between 25 percent and 75 percent presence for herbivores, and 25 percent and 75 percent emergence for plants. The species-specific indices of phenological overlap between each herbivore species and the plant community are depicted as time series for the period covered by the data in figure 7.9a. While there is a significant trend toward increasing phenological overlap between the timing of muskox arrival and plant emergence ($r = 0.59$, $P = 0.04$), there is no such trend for phenological overlap between caribou and plants ($r = -0.08$, $P = 0.80$). A t-test of mean pairwise differences in the annual extent of phenological overlap between herbivore arrival and plant emergence onset reveals that overlap is consistently and significantly greater for muskoxen (5.34 ± 0.97 days) than for caribou (2.38 ± 0.74 days; $t = -2.91$, df $= 11$, $P = 0.014$).

Moreover, analyses of the extent of phenological overlap with the timing of plant emergence reveal interesting differences between the two species of herbivores. For caribou, the single significant predictor of phenological overlap with plants is the annual date of 95 percent emergence in the plant level of the community (figure 7.9b; $r = 0.62$, $P = 0.03$). This relationship suggests that a later annual termination of the season of plant emergence results in greater overlap in time between caribou and the plant level of the system. In agreement with the pattern evident in figure 7.7 suggesting that advancing plant phenology tends to outpace the arrival of caribou at the site, an earlier termination of the season of plant emergence tends to be associated with reduced phenological overlap between caribou and plants (figure 7.9b). For muskoxen, in contrast, the extent of phenological overlap with the plant level of the community is related solely and significantly to the timing of arrival of muskoxen at the site (figure 7.9c; $r = -0.82$, $P = 0.001$). This relationship indicates that earlier arrival by muskoxen increases, perhaps not surprisingly, phenological overlap with plants. The most informative insight deriving from these analyses, however, is the distinct difference between the species in factors influencing their overlap in time with resource phenology. In caribou, this overlap is apparently driven by the phenology of the resource base itself, a further manifestation of the phenological cascade concept. In muskoxen, however, this overlap is apparently determined by phenological plasticity in muskoxen themselves, and is thus indicative of resource tracking in time by consumer phenology.

ADDITIONAL EXAMPLES OF CHANGES IN THE USE OF TIME IN TRITROPHIC-LEVEL SYSTEMS

Let us now turn our attention to additional examples of differential rates of phenological change across trophic levels in other systems. One of the most taxonomically comprehensive assessments of trophic-level phenological dynamics to date was undertaken on the basis of phenological trend data for 726 taxa compiled throughout the United Kingdom (Thackeray et al. 2010). This analysis compared 25,532 slope estimates of trends in key life history events, including dates of first flight or appearance of aphids and butterflies, laying dates and arrival dates of birds, bloom dates of freshwater and marine plankton, catch dates of moths, and first flowering dates of plants, in terrestrial, freshwater, and marine communities. Time series spanned 30 years, from 1976 through 2005, and included three trophic levels.

Across all taxonomic groups and environmental types, the mean rate of phenological change was approximately −0.04 days per decade, and nearly 84

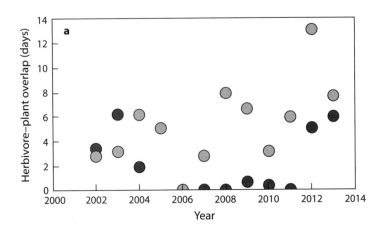

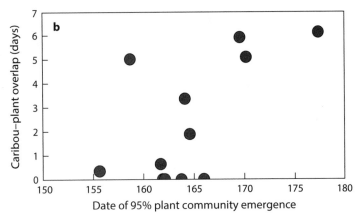

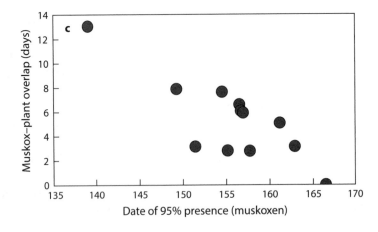

percent of the time series displayed negative trends—that is, advancing phenology (Thackeray et al. 2010). Thackeray et al. (2010) documented, however, considerable variation across taxonomic groups (but not environments) in rates and directionality of phenological change. For instance, the greatest rates of phenological advance were detected in terrestrial plants (for which the mean rate was approximately −0.06 days per decade), while the lowest were detected in freshwater plants (for which the mean rate was approximately −0.02 days per decade). As well, there was considerable decadal variation in rates of change, with low negative or positive trends, the latter indicative of phenological delays, characterizing the decade before 1985, and stronger and solely negative trends from 1986 onward (Thackeray et al. 2010).

Of greatest relevance to the discussion in this chapter, however, were the results of analyses focusing on trophic level differences in rates of change. Averaged across all taxa, environments, and decades, rates of phenological advance were stronger in primary producers and primary consumers than in secondary consumers, but did not differ between primary producers and primary consumers (Thackeray et al. 2010). Trophic-level variation was accentuated when the trend estimates were binned by decade, with rates in primary producers far outpacing

---

FIGURE 7.9. (a) Interannual variation in the degree of phenological overlap between primary consumers (caribou, black dots; muskoxen, gray dots) and forage resources (plants) at the study site near Kangerlussuaq, Greenland, from 2002 through 2013. The degree of phenological overlap in each year of observation is calculated based on the timing and duration of herbivore presence (arrival) at the site and the timing and duration of plant emergence at the site, as in the heuristic model in figure 7.1. Hence, overlap in this instance relates solely to the resource phenophase of emergence, the first phase in the annual season of primary productivity, and is calculated using estimates of phenological dispersion for each herbivore and for the plant community displayed in figure 7.7b. Years with zero or near-zero overlap are those in which dates of 25 percent presence at the site by herbivores occurred after, or nearly after, respectively, 75 percent of plant species had achieved emergence. The correlation is not significant for caribou ($r = -0.08$, $P > 0.05$), but it is significant for muskoxen ($r = 0.59$, $P = 0.05$), indicating a trend toward increasing consumer-resource phenological overlap in the latter. Factors associated with interannual variation in the degree of phenological overlap with resources differ between (b) caribou and (c) muskoxen. For caribou, there is a significant association with the timing of the conclusion of the annual period of emergence, as indexed by the annual date of 95 percent emergence at the plant community level ($r = 0.62$, $P < 0.05$). This association suggests that the degree of phenological overlap between caribou and plant resources increases as the end of the season of plant emergence is delayed, and is hence resource driven. For muskoxen, there is a significant association with the annual date of 50 percent presence of muskoxen at the site ($r = -0.82$, $P < 0.05$). This association suggests that the extent to which muskoxen overlap phenologically with plant resources increases with earlier arrival of muskoxen at the site, and is hence, in contrast to the case with caribou, consumer driven.

those of primary consumers and secondary consumers from 1986 to 1995 and from 1996 to 2005 (Thackeray et al. 2010). Such patterns are highly suggestive of the potential for the development of trophic mismatches between plants, herbivores, and predators if differential rates of phenological advance across trophic levels persist with future warming (*sensu* Both et al. 2009).

Phenological dynamics in top consumers may, however, display variation in concert with resource supply as well as abiotic environmental conditions. A long-term study of breeding phenology in tawny owls (*Strix aluco*) in Finland, for example, conducted on the basis of nest-box surveys between 1986 and 2011, detected a pronounced difference in hatching dates between urban and rural environments (Solonen 2014). The timing of hatching in urban environments was characterized by a median date of approximately April 19, while that in rural environments occurred around May 1, a difference of 12 days (Solonen 2014). Earlier hatching in urban environments was associated with greater autumn abundance there of the primary prey of tawny owls, arvicoline voles, and with milder climatic conditions during winter (Solonen 2014).

While this example indicates that top consumer phenology can be responsive to variation in resource abundance, additional evidence suggests that top consumer abundance may be vulnerable to variation in resource phenology. In tropical forest systems, it has been estimated on the basis of biomass that the majority of vertebrate taxa are specialists on flower, nectar, or fruit consumption (Fleming et al. 1987). In such systems, the timing of phenological events including flowering and fruit production by trees is triggered by stochastic precipitation events, such as those related to the El Niño Southern Oscillation, and can be delayed by drought (Shukla and Ramakrishnan 1982; Bullock and Solismagallanes 1990; Post et al. 1999a; Law et al. 2000; Numata et al. 2003; Brando et al. 2006; Brearley et al. 2007). In turn, seasonal and interannual variation in the phenology of flowering, fruiting, and nectar production can affect reproduction, survival, and abundance in an array of tropical vertebrates, including lemurs (Dunham et al. 2011), monkeys (Peres 1994; Wiederholt and Post 2010), marsupials (Wooller et al. 1993), pigs (Curran and Leighton 2000; Wong et al. 2005), birds (Corlett and LaFrankie 1998), and bears (Fredriksson et al. 2006, 2007). Increasing durations of intervals between precipitation events necessary to trigger flowering and fruiting in tropical trees, an expected consequence of continued climatic warming, thus has the potential to adversely affect a wide variety of tropical fauna through phenological cascades (Butt et al. 2015), especially if confounded by ongoing or intensifying human land use (Brodie et al. 2010, 2012; Post and Brodie 2015).

Evidence of adverse consequences for primary consumers of temporal mismatches with variation in resource phenology is abundant at higher latitudes as well, especially in pollinators (Høye et al. 2013; Kudo and Ida 2013; Kudo

2014; Petanidou et al. 2014; Solga et al. 2014; Rafferty et al. 2015; Post and Avery in press). One simulation study based on empirically derived interaction networks involving 1420 pollinator species and 429 plant species, for instance, suggests that warming owing to a doubling of atmospheric CO_2 concentration will advance flowering phenology to such an extent that between 17 and 50 percent of pollinators will suffer floral resource diminishment (Memmott et al. 2007). Consequences of such mismatch for pollinators, derived from observational and experimental studies (Forrest 2015), will be examined in greater detail later.

FINAL CONSIDERATIONS OF THE USE OF TIME IN POLLINATOR-PLANT ASSOCIATIONS

In the previous chapter, we explored consequences of differential, species-specific rates of plant phenological advance, in response to climate change, for intra- and interspecific phenological overlap and competition in time. Many elements of interspecific competition can be mediated by phenological overlap, including competition for soil nutrients, light, and water. As well, competition for pollinators may be of considerable importance in the evolution of species-specific flowering strategies and a constraint on phenological plasticity in response to climate change. Early studies of interspecific temporal dispersion, or its inverse, temporal aggregation, of flowering phenology in plant communities provide relevant insights. While abiotic factors such as the timing of rainfall in tropical systems and temperature in Arctic systems limit the duration of the community-wide flowering season, segregation among plant species in the seasonal timing of flowering is a strategy aimed at minimizing competition for pollinators (Janzen 1967; Hocking 1968).

An early, 4-year observational study of phenological dynamics in the timing of flowering by ten species of plants pollinated by hermit hummingbirds in Costa Rica (Stiles 1977) illustrates this. Between 1971 and 1974, the annual onset, peak, and end of flowering by these ten, co-occurring species were tracked, and variation in them in association with monthly rainfall assessed descriptively. Interannual variation in monthly total rainfall was matched by interannual variation in the rank order of flowering among species as well as species-specific durations of flowering (Stiles 1977). Nonetheless, Stiles (1977) concluded that, despite such variation, the dispersion in time among species remained uniform, and highlighted the adaptive value of minimal phenological overlap for reducing competition for hummingbird pollinators. A subsequent quantitative reexamination of the data, however, concluded that flowering times in this community were significantly aggregated during dry periods, rather than randomly dispersed (Poole and

Rathcke 1979). This conclusion accords with Janzen's (1967) observation that flowering times of trees in lowland Central America tend to be aggregated during the dry season.

If flowering plants are in competition for pollinators, and if the timing of flowering and temporal overlap in flowering among species within communities influences such competition, then what are the implications for performance in both flowering plants and pollinators of phenological dynamics in each? In many biomes, co-occurring insect-pollinated plants display considerable interannual variation in intra- and interspecific temporal dispersion of flowering dates that appear driven by abiotic conditions. For instance, in a temperate North American community, flowering times of 14 co-occurring shrub species overlap almost completely during late-flowering years but much less so during early-flowering years (Rathcke 1988). Interspecific competition for pollinators is, in this instance, apparently mediated by pollinator diversity (Rathcke 1988). In a community of *Acacia* sp. trees in northern Tanzania, flowering is highly constrained temporally and therefore characterized by considerable overlap among species (Stone et al. 1996). In this system, competition for pollinators is minimized by fine-scale partitioning of time: the timing of peak pollen release in this assemblage varies on an hourly basis among *Acacia* species (Stone et al. 1996). While such examples highlight phenological strategies plants employ to maintain pollination services, others illustrate adverse consequences for plants of loss of pollinator services, and for pollinators of loss of floral resources, resulting from phenological mismatch. Such mismatch occurs when phenological advancement in response to the alleviation of abiotic constraints on the annual timing of flowering is not sufficiently tracked by phenological advance in pollinators.

For instance, consider the results of an analysis of the consequences of differential plant and pollinator responses to observed climate change over a 10- to 14-year period in Japan (Kudo and Ida 2013). The flowering phenology of a forest floor species in the poppy family, *Corydalis ambigua*, was monitored concurrently with the timing of activity of its bumblebee pollinators, primarily *Bombus hypocrita*, at three sites in Hokkaido, Japan, from 1999 to 2013. Across all three sites, timing of flowering advanced with earlier snow melt and warmer spring temperatures, and did so at a greater rate than that of bumblebee activity (Kudo and Ida 2013). Hence, earlier flowering resulted in greater phenological mismatch between *C. ambigua* and its pollinators. As a consequence, seed production declined in *C. ambigua* (Kudo and Ida 2013). In a related study, also conducted in Hokkaido, Japan, the flowering phenologies of 14 fellfield and snowbed species were compared between two climatically divergent seasons in 2011 and 2012 (Kudo 2014). In the latter year, which was characterized by an unusually warm spring, flowering times of all 14 species in each system advanced. Within the pollinator community, however, only one species of bumblebee advanced its timing of activity in pace with advanced flowering (Kudo 2014).

Results of experimental manipulation of flowering phenology appear to agree with those of the long-term observational study in Hokkaido, indicating reduced reproductive success of plants as a consequence of phenological mismatch with pollinators. For example, a greenhouse treatment was used to advance the flowering times in two species of wildflower common to the upper Midwest of the United States, after which the plants were placed in a natural outdoor environment with control plants and exposed to insect pollinators (Rafferty and Ives 2012). For one of the species, experimentally advanced flowering exposed it to a compositionally different pollinator community compared to control plants. The same applied to the other species, but it also experienced reduced pollinator effectiveness as a consequence of experimentally advanced flowering. These results suggest that phenological mismatch with pollinators as a consequence of climate change may reduce reproductive success of flowering plants (Rafferty and Ives 2012). A similar experiment conducted in the same system, but involving 14 species of flowering plants, revealed divergent consequences of advanced flowering phenology for early- versus late-flowering species (Rafferty and Ives 2011). Plant species characterized by early-season life history strategies experienced increased pollinator visits in response to experimentally advanced flowering phenology. In contrast, those species characterized by later-season life history strategies experienced reduced pollinator visits after their flowering phenology was experimentally advanced (Rafferty and Ives 2011). These results suggest that advancement of flowering phenology in response to climate change may, in some species, be constrained by co-evolutionary relationships with pollinators (Rafferty and Ives 2011).

Last, consider the consequences of variation in phenological overlap between plants and pollinators in a system characterized by rapidly shifting timing of seasonal activity. The greatest rates of phenological advance represented in the data set used in the meta-analysis in chapter 2 have occurred at the High Arctic site Zackenberg, in Greenland (Forchhammer et al. 2005; Høye et al. 2007; Post et al. 2009; Iler et al. 2013b). At this site, the pollinator guild comprises butterflies, muscid flies, and midges. Among butterflies, the emergence phenology of two species was analyzed for variation in association with abiotic factors and the flowering phenology of their floral resources between 1996 and 2009 (Høye et al. 2014). Despite the fact that the mean, start, peak, and termination of the annual flight season differed between the two butterfly species, and were differentially responsive to abiotic drivers, the duration of the flight season in both was positively correlated with the duration of flowering by their host plants (Høye et al. 2014). Hence, there appears to be relatively tight coupling between the phenological dynamics of pollinators and plants, despite trends toward earlier flowering.

In contrast, earlier and shorter flowering at the same site between 1996 and 2006 reduced temporal overlap between community-wide flowering and muscid and midge activity (Høye et al. 2013). As a consequence, abundance of both pollinators declined at the study site by approximately two-thirds in one decade in association with reduced temporal overlap with floral resource phenology (Høye et al. 2013). This series of associations, from growing season warming, to earlier and more abbreviated community-level flowering, to reduced phenological overlap among pollinators and flowering plants, to, last, declining abundance of pollinators, is illustrative of an indirect phenological cascade.

What insights have we derived from these examples? The allocation of time by individuals to life history events influences intraspecific overlap in time, interspecific overlap in time, and the phenology of trophic interactions. Similarly, responses by individuals to changes in the availability of absolute and relative ecological time shape not only phenological communities, but also interactions that scale from individuals to communities. However, with some exceptions, most of the insights up to this point have been drawn from consideration of dynamics in highly seasonal environments. We may wonder how well the main elements of this framework for thinking about time in ecology apply to or predict phenological dynamics in less seasonal environments such as the tropics. The next chapter will address potential conceptual shortcomings of the framework while highlighting examples from tropical systems that do in fact appear to accord with its predictions.

CHAPTER EIGHT

Limitations and Extension
to Tropical Systems

Development of this theoretical framework has relied extensively on the study of phenology in moderate to highly seasonal environments. In such environments, time is limited. By this is meant the availability of time an organism can allocate to growth, maintenance, and offspring production is limited on an annual basis. This implies that the degree of time limitation on an annual basis increases with latitude. Conversely, closer to the equator, the availability of such ecological time—that is, the amount of time available for biological production on an annual basis—should be comparatively abundant. It might even be argued that ecological time should be nearly unlimited in tropical systems because there photoperiod and temperature are favorable for growth and offspring production throughout the year (van Schaik et al. 1993).

The explanation for this latitudinal gradient in the availability of ecological time lies in the increase in seasonality, defined as intra-annual variation in diurnal photoperiod and temperature, with latitude that is shown earlier in figure 2.1. To this, we can add latitudinal variation in the amount of solar insolation throughout the year (figure 8.1a). As with day length or diurnal photoperiod, there is comparatively little variation throughout the year in incoming solar radiation at the equator, but considerable variation at the poles. This is not to say that solar irradiance is nearly constant throughout the year in the tropics for, indeed, it varies substantially on a seasonal basis with cloudiness and atmospheric moisture content (van Schaik et al. 1993). Nonetheless, whereas the equator receives approximately 35 megajoules of solar radiation per square meter every day throughout the year, solar insolation at the poles varies from approximately zero during polar winter to 45 to 50 megajoules per square meter per day during polar summer (figure 8.1b). This necessitates increasing rates of change, both positive and negative, in incoming solar radiation with latitude across the annual cycle. Hence, the availability of ecological time for growth, maintenance, and offspring production, while more limited on an annual basis as one moves away from the equator, increases and decreases more rapidly on a subannual basis along the same gradient.

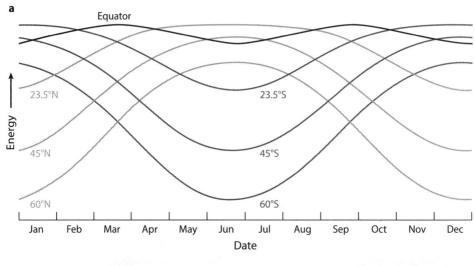

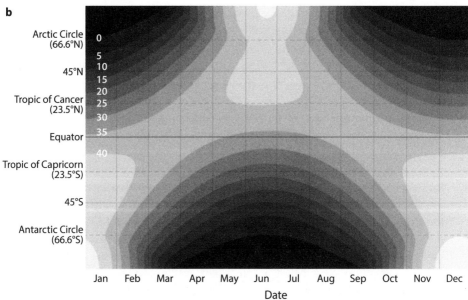

FIGURE 8.1. Monthly variation in total solar energy input to the surface of the Earth in relation to latitude. In panel (a), the magnitude of variation is not quantified along the *y*-axis, but is depicted in a relative manner. In panel (b), total solar energy input is represented by shading in relation to megajoules of energy. Adapted from "Climate and Earth's Energy Budget," by the NASA Earth Observatory, http://earthobservatory.nasa.gov/Features/EnergyBalance/page3.php, original illustration credit Robert Simmon.

Yet organisms in both types of environments, highly seasonal and highly aseasonal, exhibit strongly dynamic phenologies. What, then, drives variation in the use of time by organisms in such widely different types of environments? In other words, how generally would an ecological theory of time apply to phenology or life history dynamics in less seasonal environments such as the tropics? Any sound theory must generate testable predictions (Kuhn 1962). If it applies generally, then, the theoretical framework developed thus far and predictions generated by it should apply to phenological dynamics in tropical systems as well as in high-latitude systems.

It may turn out that extension of the framework to tropical systems will provide us with testable predictions relating to increases in the availability of ecological time. This is because as we move from the poles to the tropics, limits on the availability of ecological time become relaxed. There are three main predictions where we can look for applicability of the framework to tropical systems. First, given more ecological time, organisms should use it. This should be apparent in two ways. Species should display evidence of phenological activity related to growth, maintenance, or offspring production throughout a greater portion of the year in tropical systems than in higher latitude systems. As well, factors constraining phenophase duration should vary from low to high latitudes, but if such constraints are primarily abiotic, then the allocation of time to any given phenological stage should be greater for that phenophase in the tropics than for the same phenophase at higher latitudes. Flowering times, for instance, of species or genera whose distributions span tropical and extra-tropical regions should be longer in the former compared to the latter. This prediction night not apply, however, if competition replaces abiotic limitation on phenophase duration as one moves from high- to low-latitude environments because species richness increases dramatically along the same gradient. Second, with more ecological time, species should demonstrate the capacity for segregation in time from closely related species in the same assemblage. This relates to the use of time as a buffer. Last, phenological dynamics in the tropics should be consistent with the notion of relative ecological time, displaying temporal overlap with resource species or strategies for the use of time that limit exposure in time to consumers. This relates to the use of time as a refuge.

Let us now examine each of these predictions in turn, using two approaches. The first will be an examination of patterns of phenological dynamics in tropical systems and the factors that explain them. The second will include examination of whether greater availability of ecological time in tropical systems compared to higher latitude systems is reflected in patterns of the use of time within species and among species in horizontally and vertically structured phenological communities.

SEASONAL AND INTERANNUAL PHENOLOGICAL
DYNAMICS IN TROPICAL SYSTEMS

Although abiotic conditions are conducive to growth and reproduction through-out the year in the tropics, many tropical species exhibit pronounced phenological dynamics that shape and are influenced by their interactions in time with other spe-cies (Butt et al. 2015). Hence, even though ecological time is abundant in tropical systems, species can still be seen to segregate their phenological activity in time. This is evident in the leaf flushing, flowering, and fruiting activity of many tropical plant species that, while capable of expression of such life history events throughout the year, tend to engage in peaks in such activity that lend structure to the phenologi-cal community (van Schaik et al. 1993). In contrast to high-latitude systems, where community-level leaf flushing peaks over a period of days, such dynamics can be prolonged yet still periodic in the tropics. Annual peaks in the number of species undergoing leaf flushing, flowering, fruiting, or abscission can, in tropical moist forests, last several weeks (Shukla and Ramakrishnan 1982). The annual period of new leaf production by the canopy tree *Manilkara hexandra* in tropical forests of Sri Lanka, for instance, is timed to coincide with the annual rainy season and typically lasts eight to ten weeks (Gunarathne and Perera 2014). Hence, plants in tropical systems allocate time to growth and reproduction in a highly variable man-ner on sub- and superannual scales, even though variation in temperature and light availability on a diurnal basis may exceed that on an annual or interannual basis in such systems (Wright 1991). In fact, phenological variation, coinciding with abiotic seasonality of one form or another, as will be examined next, appears to be the norm rather than the exception among tropical species (van Schaik et al. 1993).

A major review of phenological dynamics in tropical forest species revealed that the majority of woody deciduous species display periodic rather than con-tinuous production of new leaves and flowers (van Schaik et al. 1993). Moreover, at the community level, the seasonal timing of peaks in the number of species producing new leaves or flowers tends to vary latitudinally within the tropical zone across Africa, Asia, and the Americas (van Schaik et al. 1993; see figure 8.2a). This variation appears to track the timing of seasonal solar irradiance. This suggests that seasonal variation in incoming solar radiation or light availability is a major driver of the phenology of leaf and flower production in many tropical plant species, although this is likely mediated proximally in tropical dry forests by moisture availability (van Schaik et al. 1993). The annual timing of fruit produc-tion by fleshy-fruited species also varies with latitude both globally and within the tropics in association with several environmental variables, including precipita-tion, temperature, and solar energy input (Ting et al. 2008; see figure 8.2b). Simi-larly, the duration of the annual season of fruit production in fleshy-fruited species is greatest at the tropics and declines poleward from there in both hemispheres in association with variation in actual evapotranspiration (Ting et al. 2008; see figure 8.2c).

Further, in moist tropical forests, cloud cover and shading induce seasonal photo-limitation of net primary production (Loescher et al. 2003). In such systems, flowering and fruiting tend to peak during the dry season, when solar irradiance is also at its annual peak (van Schaik et al. 1993). On Barro Colorado Island, Panama, for instance, this is evident in the annual peak, during the months of March and April, in numbers of species engaged in flowering or fruiting, which can be nearly twice as great as the numbers of such species engaged in reproductive phenology during the wet season (Wright and Calderon 2006). Conversely, flowering in the canopy tree *M. hexandra* in Sri Lankan deciduous forests can be initiated following a decline in solar radiation if this is accompanied by reduced air and soil temperatures (Gunarathne and Perera 2014).

Nonetheless, identification of general drivers of phenological dynamics in tropical systems has been elusive, further emphasizing the idiosyncratic nature of life history strategies relating to the use of time within and among species. A comparison of factors influencing the flowering phenology of four eucalypt species over a 23-year period in Australia illustrates this point. Timing of flowering in two of the four species varied with local temperature, while an interaction between rainfall and temperature predicted the timing of flowering in all four species (Keatley et al. 2002). However, this latter relationship indicates that increases in warming and moisture availability should advance the timing of flowering in two of these species and delay it in the other two (Keatley et al. 2002). This is consistent with expansion into increasingly available ecological time, and increased phenological dispersion among species, by high-latitude plants documented in chapter 6. Similarly, another analysis of flowering phenology of four closely related species of Australian eucalypts revealed substantial variation among species in the annual duration of flowering as well as idiosyncratic responses of flowering timing among these species to maximum, minimum, and mean monthly temperature (Hudson et al. 2010). An examination of factors influencing variation in the timing of flowering among 20 eucalypt species along the coast of New South Wales, Australia, revealed that flowering in 9 species was cued by cooling, while the highest degree of flowering across all 20 species occurred 9 months after heavy rainfall (Law et al. 2000). The implications of such species-specific responses to the alteration of abiotic conditions that constrain the availability of ecological time will be discussed in the context of competition in horizontally structured communities later in this chapter.

The timing of different phenophases may likewise be cued by different factors. Among 122 tree species observed in a humid subtropical forest in northeastern

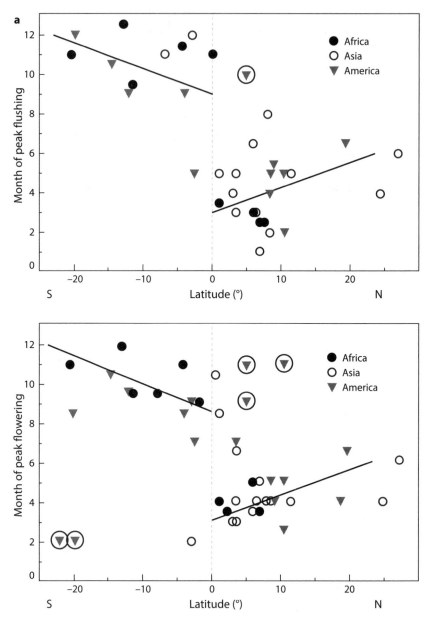

FIGURE 8.2. (a) Even within the tropical zone, latitudinal gradients in phenological timing are evident. In the top panel, the timing of peak leaf flushing (month of the year) varies with latitude across the tropics. The bottom panel shows the same pattern for timing of peak flowering. Adapted from van Schaik et al. (1993). (b) Latitudinal variation in the peak timing of fruit

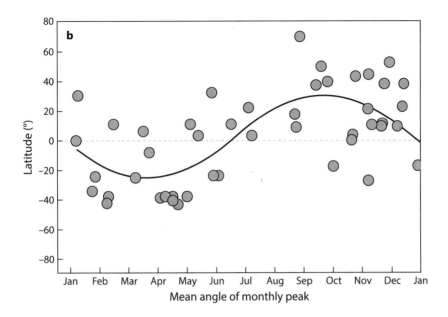

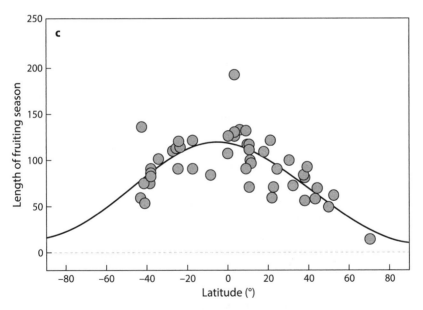

production by fleshy-fruited species of plants across the Earth in a global assessment of fruiting phenology. Here, peak timing (month) is quantified as the mean angle derived from a circular annual calendar. (c) Latitudinal variation in the length of the fruiting season by fleshy-fruited species derived from the same assessment. Adapted from Ting et al. (2008).

India, loss of leaves coincided with the onset of the dry season, formation of flower buds coincided with the peak period of leaflessness during the dry season, while leaf flushing and blooming tended to coincide with and peak during the rainy season (Shukla and Ramakrishnan 1982). Experimental evidence further suggests that moisture availability may be important in influencing the timing of leaf flushing and flowering in tropical moist forests. Experimental irrigation of plots on Barro Colorado Island, Panama, for instance, altered the association between the occurrence of the wet season and the timing of new leaf production in all seven species examined (Wright 1991). Similarly, experimental drought inducement delayed the timing of leaf formation, flowering, and fruiting of a focal tree species, *Coussarea racemosa*, in the Tapajos National Forest, Brazil (Brando et al. 2006). Because abiotic factors such as temperature, moisture availability, and solar irradiance tend to be highly correlated, identifying primary drivers of phenological dynamics in tropical species may be facilitated by insights from studies of longer-term variation.

A 10-year study of the timing of flowering by 171 species of trees in a tropical wet forest in central Borneo revealed what might best be described as remarkably stochastic long-term variation in reproductive phenology (Brearley et al. 2007). Over this 10-year period, between 1990 and 2000, the authors observed, on the basis of monthly surveys, only three mass flowering events. The vast majority of species, 73 percent, flowered irregularly and on a supra-annual basis. Interestingly, in contrast to shifts in the rank order of emergence among species at the Low Arctic study site in Kangerlussuaq, Greenland (Post et al. 2016), sequential patterns of flowering by species in three genera of dipterocarps at this study site in Borneo were largely consistent in each of these three mass flowering events (Brearley et al. 2007). This observation accords with the hypothesis that closely related species may partition the use of time allocated to flowering as a means of minimizing competition for pollinator services (Brearley et al. 2007). While the authors were in this instance unable to clearly isolate individual abiotic drivers of irregular supra-annual flowering in this system, the three flowering events appeared to coincide with moisture limitation associated with El Niño Southern Oscillation (ENSO) dynamics (Brearley et al. 2007). Similarly, mass flowering events on the Malay Peninsula appear to be triggered by or strongly associated with ENSO dynamics. Observations over a 22-year period between 1980 and 2002 revealed the occurrence of 11 mass flowering events in a lowland moist tropical forest associated with seasonal moisture limitation and cooling that may have been ENSO-driven (Numata et al. 2003).

Evidence from a moist tropical forest on Barro Colorado Island largely corroborates these associations, but provides an alternative interpretation for them. At the community level, seasonal flowering at this site peaks in March and April, while seasonal fruit production peaks in April; both phenophases coincide with the end of the annual dry season (Wright and Calderon 2006). Longer-term variation over an 18-year period in patterns of flower and seed production indicates, in general agreement with the preceding description of long-term flowering dynamics on Borneo, an association with the occurrence of moderate El Niño events (Wright and Calderon 2006). In this case, however, Wright and Calderón (2006) argue that this is not a response to moisture limitation itself, but rather a response to increased solar irradiance associated with ENSO dynamics because net primary production in the humid tropics, they argue, tends to be limited by light availability. Though not strictly focused on phenological dynamics, a more recent analysis of long-term trends in flower production among multiple species at the community level in a seasonally dry forest on Barro Colorado Island indicates that such reproductive activity is most closely associated with multiannual variation in temperature (Pau et al. 2013). At this site, seasonal variation in flower production was associated with an interaction between cloudiness and temperature (Pau et al. 2013). However, a significant long-term increase in total flower production between 1987 and 2009 at the site was associated with a modest increase in maximum annual temperature, but was unrelated to interannual variation in cloudiness (Pau et al. 2013). These results further emphasize that identification of primary abiotic drivers of phenological dynamics in tropical systems may be complicated by interactions between seasonal and interannual dynamics. Perhaps more importantly, however, they indicate the potential for considerable interannual variation and even trends in reproductive phenology in tropical species.

Just as long-term observations on Barro Colorado Island indicate increases in flower production at the community level associated with increased temperature, long-term phenology records from a tropical moist forest in Kibale National Park, Uganda, indicate trends toward increased fruiting (Chapman et al. 2005). Between 1990 and 2002, the percentage of trees engaged in fruiting has increased nearly tenfold (Chapman et al. 2005). At the same site, rainfall has increased by approximately 300 millimeters over the past century, droughts have declined in frequency, and monthly maximum temperatures have increased by 3.5° C over the past 25 years (Chapman et al. 2005). Nonetheless, there was considerable variation among four sites within the park in the association between rainfall and fruiting phenology (Chapman et al. 2005).

On a global basis, examination of long-term trends in primary production, indexed by satellite-derived Normalized Difference Vegetation Index (NDVI) data, has revealed important insights into phenological variation over decadal scales in the tropics. An analysis of multidecadal Advanced Very High Resolution Radiometer (AVHRR) data revealed, for example, detectable intra- and interannual variability in vegetation production across the tropics, including regions in South America, sub-Saharan Africa, Southeast Asia, and northern Australia

(Garonna et al. 2016). Moreover, this analysis revealed trends toward increasing growing season length across much of the same band, but primarily in sub-Saharan Africa, due mainly to trends toward delays in end-of-season primary production (Garonna et al. 2016). In fact, approximately 12 percent of extremely warm and moist tropical forest globally has undergone a significant increase in growing season length over the past three decades, ranking this biome fourth among 16 vegetation zones in seven biomes (Garonna et al. 2016). Furthermore, a comparison among significant trends only reveals that tropical biomes have undergone the greatest average increase in growing season length since 1982 (Garonna et al. 2016). Over the instrumental record, however, the rate and magnitude of tropical warming have been considerably less than at higher latitudes (IPCC 2014).

PARTITIONING OF TIME IN PHENOLOGICAL COMMUNITIES IN THE TROPICS

Chapter 6 presented a conceptual model of the partitioning of time by species in horizontally structured phenological communities. This model was intended to explain, and to predict changes in, the degree of temporal overlap within and among species in their respective timing of life history events related to growth and offspring production. According to this model, as the availability of ecological time expands or contracts we should expect to see, respectively, increased or reduced segregation in time among individuals within species and among species themselves. Multiannual data on the timing of annual growth onset by plants at the Kangerlussuaq study site in Low Arctic Greenland appear to accord with these predictions: in years with earlier onset of growth, phenological dispersion increases within species while phenological segregation among species also increases (see figure 6.6).

If we apply the predictions of this model to tropical systems, what might we expect to see and what does the evidence indicate? As already suggested, ecological time must be in greater abundance on an annual basis in the tropics than in higher latitude systems. Hence, we should expect to see, in tropical systems as compared to higher latitude systems, both increased phenological dispersion within species and segregation among species across a longer annual period—that is, an infilling of available ecological time.

Patterns of flowering phenology by myrtaceous trees in New South Wales, Australia, documented in the long-term observational study by Law et al. (2000) described in the previous section are consistent with both of these expectations. First, flowering by trees in the five genera comprising these 20 species occurs throughout the year (figure 8.3). In contrast, community-wide flowering in the

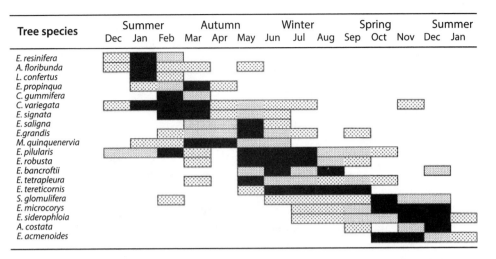

FIGURE 8.3. Use of annually available ecological time by, phenological dispersion in, and phenological segregation among 20 species of myrtaceous trees in tropical Australia. Shown are annual periods of flowering by each species. Light shading indicates periods of occasional flowering, gray indicates periods of regular flowering, and black indicates periods of most intense flowering. Adapted from Law et al. (2000).

High Arctic may be as short as 19 days (Høye et al. 2013). Second, each individual species engages in a period of flowering that can last from a few to several months (figure 8.3). This is indicative of a remarkable degree of intraspecific phenological dispersion that dwarfs the degree of dispersion seen in species at the high-latitude Kangerlussuaq field site (see figure 6.6). Despite this, there is a clear tendency toward segregation in time among the 20 myrtaceous species in this tropical system. Annual peaks of species-specific flowering are largely nonoverlapping across the community (figure 8.3, black bars). Hence, most species in this system displayed a distinct flowering season, even though this could be prolonged (Law et al. 2000). This suggests that the use of time as a buffer against interference interactions (*sensu* chapter 6) applies in tropical systems as well as at higher latitudes.

Such segregation in time among species in tropical systems may also reflect a strategy aimed at reducing competition for interactions with consumer species such as pollinators, nectarivores, or frugivores. For instance, largely nonoverlapping flowering times of congeneric species of three genera of trees in subtropical India, *Eugenia, Garcinia*, and *Litsea* spp., may minimize competition among closely related species for pollinator services (Shukla and Ramakrishnan 1982). In chapter 7, a conceptual model was presented that was intended to describe and predict changes in the use of time in such vertically structured communities. According to that model, consumers may be expected to adjust the seasonal timing of expression of life history traits to track shifts in the phenology of resource species. Conversely, resource species may be expected to shift the timing of expression of life history traits to minimize or escape exploitation by consumers. In the special case of pollinator-plant interactions, however, species at both trophic levels may be viewed as exploitative of, and in fact dependent for their own reproductive success upon, the phenology of species at the other trophic level. Hence, in this case, any shift in the phenology of plant species in response to a change in the availability of ecological time should be tempered by counterselection for matching with pollinator resources. Barring such mediating counterselection, shifts in plant phenology should be closely matched by shifts in pollinator phenology.

In tropical systems, such dynamics apply as well to the seasonal timing of activity by frugivorous consumers, which can be important agents of seed dispersal (Butt et al. 2015) or which can, conversely, limit the reproductive success of species they consume (Butt et al. 2015). As noted earlier, van Schaik et al. (1993), provide a comprehensive review of phenological segregation and aggregation among tropical forest plant species, and interpret these strategies according to their adaptive value under different selective factors. Their insights are applicable here. Generally, within tropical plant communities, segregation among species in their timing of flowering may best be understood as an adaptation to minimize competition for pollinators (van Schaik et al. 1993). Phenological aggregation in time may, however, be adaptive under a variety of scenarios relating to the use of relative ecological time as predicted by the model for exploitation interactions in chapter 7. Aggregated timing of flowering by tropical species may reflect selection for the timing of such activity to coincide with the seasonally limited presence of pollinators (van Schaik et al. 1993). Aggregated timing of fruit production by tropical species may, alternatively, reflect selection for the avoidance or minimization of losses to consumers, which accords with the notion of the use of time as a refuge (figure 7.2). This is most clearly illustrated in masting behavior common to many tropical tree species (van Schaik et al. 1993). Hence, while biotic factors related to consumer-resource interactions may influence the degree of segregation or aggregation of phenological activity in time, abiotic factors such as those reviewed in the previous section determine the timing of the peak in phenological aggregation (van Schaik et al. 1993).

AVAILABILITY OF TIME RELATED TO ENERGY INPUT

While photoperiod, and seasonal variation in diurnal light:dark ratios, is an important driver of phenological dynamics in mid- and high-latitude systems, it is not in tropical systems, where photoperiod is largely invariant throughout the year. This raises the interesting observation that life has evolved differential sensitivity to, and responsiveness to, environmental cues indicative of the availability of time for allocation to biological processes across the Earth's latitudes. Away from the equator, these cues are clearly photoperiod and associated seasonal changes in temperature. Near the equator, in contrast, they include such factors as moisture availability and solar irradiance, the latter a direct measure of energy input.

Despite these differences, solar insolation is a key driver of biological activity and the related allocation of time to growth and offspring production. So how do we account for the different patterns of temporal activity across seasons between low- and high-latitude systems? Solar insolation is largely constant in tropical latitudes throughout the year, and greater on an annual basis, than at higher latitudes. It must, therefore, be the rate of change in solar energy input across seasons that drives the much more clearly defined and periodic phenological dynamics characteristic of higher latitude systems. Both seasonal increases and declines in solar energy input increase with latitude, while total annual solar insolation does not (figure 8.1). Hence, the *seasonal rate of change* in solar energy input dictates changes in the availability of time for allocation to biological activity at mid- and high latitudes, but not at low latitudes. At low latitudes, production and reproduction are possible throughout the year, but tend to occur mostly in pulses separated by prolonged periods. These pulses are apparently cued by interactions between biotic and abiotic factors. The former relate to competition and resource exploitation interactions with consumers reflective of the use of relative ecological time. The latter relate to climate- and weather-driven variation in the availability of moisture and light, and hence to the use of both absolute and recurrent ecological time.

After entertaining such considerations, though, we are left in the end with the conclusion that generalized differences in life history dynamics and phenology between high-latitude systems and low-latitude systems relate ultimately to differences in the seasonality of energy input to them. This conclusion suggests that organisms have evolved the necessary means for capturing and allocating available tine for growth, maintenance, and offspring production characteristic of the environments in which they occur. In the final chapter, we will explore the role of time in ecology in a more general context.

CHAPTER NINE

The More General Role
of Time in Ecology

What is the role of time in ecology? Ask a colleague this question, and you are likely to hear something about the role of seasonality in determining the availability of nutrients and other resources (Tilman 1988; Shaver and Kummerow 1992); time as a niche axis along which individuals and species segregate to minimize competition (Chesson 2000; Fargione and Tilman 2005); constraints on intra- and interseasonal interactions among organisms (Shea and Chesson 2002); the tempo and scale of population dynamics and cycles (May 1974, 1975; May and Anderson 1979; Turchin and Taylor 1992; Bjørnstad and Grenfell 2001); the rate of spread of infectious diseases (Grenfell et al. 2001; Bjørnstad et al. 2002); the processes of species assemblage and succession (Chesson and Huntly 1989; Brown and Heske 1990; Harte and Shaw 1995); dispersal and range dynamics; micro-evolutionary adaptation to climate change (Skelly et al. 2007; Visser 2008); or, in the case of phenology, the onset and progression of the expression of life history traits (Lieth 1974; Mahall and Bormann 1978; Schwartz 2003). Each of these represents a well-established and thoroughly studied conceptualization of the role of time in ecology. But in each case, time is perceived as the stage upon, background against which, or metric of the rate at which, ecological processes unfold.

The argument that has been developed here, in contrast, posits that time is a central element in ecology, a resource in and of itself. This conceptualization of time attempts to explain phenological patterns and dynamics across the population, species, and community levels as consequences of the use of time, as a resource, by individual organisms. It is an attempt to make sense, in ecological and evolutionary contexts, of widely varied patterns of phenological dynamics among individuals within species, among species within communities, and among communities within and across biomes. In essence, these patterns comprise phenological advance, phenological stasis, and phenological delay, and examples of each are evident at all of these levels of biological organization and in all systems and biomes studied to date. If there is a unifying explanation for such apparently inconsistent patterns, it may well lie in the singular strategy employed by all living organisms: the conversion of time into biomass and progeny.

The introduction presented the basic elements of what is elaborated upon throughout this book as a theoretical framework for time in ecology, and chapter 3 drew distinctions among cosmological time, recurrent time, and relative ecological time. With these elements and distinctions in mind, the overarching summation of this framework is that living organisms make use of recurrent or relative ecological time to perpetuate their genes through cosmological time. And this, of course, implies challenges to the individual in contending with and capitalizing upon variation in the availability of time. Cosmological time and recurrent time unfold at predictable rates that are detectable and traceable by the individual organism, either endogenously or exogenously, as changes in photic conditions over daily, seasonal, and annual periods (Aschoff 1960, 1966; Pittendrigh 1960; Lu et al. 2010). Relative ecological time, however, may unfold in fits, spurts, and varying rates depending on sudden or gradual phenological shifts or changes or abrupt events in phenological dynamics of other components of the ecological system. Hence, cosmological and recurrent time are external to the biological system, existing and progressing even in its absence, while relative ecological time is internal and inherent to the biological system of species interactions. In this context, the increasing rate of land-surface warming at high northern hemisphere latitudes reported in chapter 2 suggests a disproportionate increase in the availability of relative ecological time in those systems for temperature-limited species than for those limited by photoperiod, the seasonal variation of which does not change with warming.

Implications for phenological mismatch by the individual organism arise from the existence of these various forms of time. Individuals may, for instance, maintain absolute time, but lose relative ecological time. Species of birds that undergo long-distance migrations may time such migrations based on abiotic cues that promote remarkably consistent arrival on breeding grounds in cosmological time, but remarkably inconsistent arrival in relative ecological time (Forchhammer et al. 2002; Both et al. 2005; Clausen and Clausen 2013). Alternatively, individuals may gain absolute cosmological time by advancing migration or reproductive phenology, but still lose relative ecological time by lagging behind even greater advances in resource species phenology (Both et al. 2009). As well, individuals can use space to acquire more time. Whereas the use of space requires use (and loss) of cosmological time, it can result in the acquisition of more relative ecological time. For instance, migration to and from breeding and nonbreeding ranges results in the maintenance of, or reduced loss of, temporal overlap with other resources and the avoidance of natural enemies (Tyler and Øritsland 1989). Similarly, seasonal elevational migrations, while consuming cosmological time, increase the availability of relative ecological time by maintaining access to forage resources that exhibit phenological waves upslope (Albon and Langvatn 1992; Hebblewhite et al. 2008). The constraints imposed upon phenological adjustment to variation in the availability of relative ecological time by life history adaptations reflecting selection in relation to photoperiodic control can, however, result in phenological mismatch (Both et al. 2009; Yang and Rudolf 2010). This mismatch, in turn, can have adverse consequences for successful offspring production (Visser et al. 2004; Both et al. 2006; Visser et al. 2006; Post and Forchhammer 2008; Kerby and Post 2013b; Plard et al. 2014); in other words, a failure by the individual organism to successfully use relative ecological time to perpetuate its genes through cosmological time.

In the simplest sense, life itself can be viewed as the use of time to these ends. Some species have evolved means of maximizing efficiency in the use of time to convert energy and nutrients to biomass and offspring, while others have evolved means of more gradual conversion of time to biomass and offspring, as early r- and K-selection theory and its conceptual predecessors proposed decades ago (Fisher 1930; Williams 1966; Gadgil and Bossert 1970; Pianka 1970; Shapiro 1975). But even within populations of a species, we can observe relative differences among individuals in their use of time. This is exemplified by, for instance, variation in rates of development, in onset of reproductive maturity, and in onset of senescence. And the same applies to variation among populations within a species across its distribution, among species within a community, among communities across biomes, and so on.

As has also been emphasized throughout this book, the individual organism's use of recurrent time and relative ecological time is constrained by three principle abiotic factors, or some combination of them: (1) photoperiod or solar irradiance; (2) temperature, or more specifically minimum annual or seasonal temperature; and (3) precipitation or moisture regime. These factors may be considered the principle constraints on, or cues triggering, the seasonal timing of expression of life history traits related to growth, development, and reproduction, and their relative magnitudes of effect in relation to one another arrayed in a two-dimensional parameter plane (Garonna 2016; see figure 9.1a). All three of these factors covary, of course, across scales of space and time. As discussed in chapter 2, this spatiotemporal co-variation manifests as patterns of seasonality across the Earth, and furthermore accounts for the distribution of biomes across the planet (Reich et al. 1997).

In an idealized setting, the timing of expression of life history traits by the individual organism would be limited solely by one of these factors. In this case, the positioning of an organism in the parameter plane quantifying the magnitude of the factor limiting its life history timing, and thus the use of time itself, will

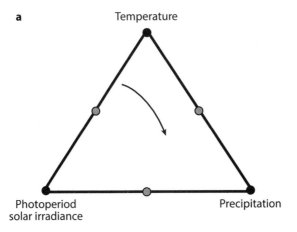

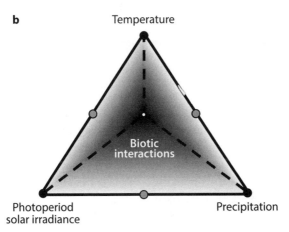

FIGURE 9.1. (a) The three principle abiotic factors limiting or cuing the timing of expression of life history traits comprising phenology as arrayed in a two-dimensional parameter plane. Modified from Garonna (2016). Coefficients quantifying their relative importance to the timing of life history traits can be plotted in the parameter plane. Absolute importance of a single factor is indicated by a coefficient occurring at one of the apices (black points). A combined influence of any two factors is indicated by a coefficient occurring along one of the axes (gray points). A combined influence of all three factors is indicated by a coefficient occurring within the interior of the parameter plane. The arc (arrow) indicates a shift in the relative importance of temperature and precipitation, with maintenance of a constant influence of photoperiod or solar irradiance. In panel (b), a fourth factor representing the influence of biotic interactions has been added, resulting in a three-dimensional parameter space. The white point indicates absolute importance of biotic interactions in limiting or cuing the timing of expression of a life history trait. Positioning on the parameter plane (a) or within the parameter space (b) may vary in several ways: among traits within individuals, if, for example, the timing of emergence, flowering, and fruit set by an individual plant are cued by different factors; within traits among individuals, if, for example, factors limiting the timing of emergence in a long-lived plant vary with climate change across its life; among species within traits, if, for instance, the factors influencing the timing of nest initiation differ among long- and short-distance migratory birds; and within or among traits across the distribution of a species.

fall at one of the endpoints of each axis (figure 9.1a, black points). More realistically, however, the timing of expression of such traits is limited by the interaction between two of the factors, in which case the positioning of the organism in the parameter plane will lie somewhere along one of the axes (figure 9.1a, gray points); or by some combination of all three factors, in which case the organism will fall somewhere inside the parameter plane (figure 9.1a, white area).

And the position of the individual within this parameter plane may, of course, vary. For instance, the position of a long-lived individual of a given species according to factors limiting the timing of expression of a given life history trait within its life cycle may shift within the parameter plane if the relative importance of each limiting factor changes from season to season across the organism's life span. The positions of individual life history traits within the parameter plane may also vary for a given species according to the factors limiting the timing of their expression. For instance, the annual timing of emergence by a plant may be cued by a combination of photoperiod and temperature, while timing of flowering in the same plant may be cued primarily by temperature, the timing of emergence, and precipitation or moisture regime (Janzen 1967). Likewise, the relative importance of each factor in cuing different life history stages or phenophases within species may shift under climate change. Such variation may also be observed among traits within species and within traits among species across spatial gradients encompassed by species distributions. Positioning in the parameter plane of traits within species or traits among species should thus be viewed as dynamic in time and space.

Indeed, recent work on global-scale patterns of variation in vegetation phenology suggests the relative importance of each of these factors in biome-wide metrics of the annual timing of the start and end of the growing season has changed over the past three decades, with precipitation/moisture limitation assuming greater proportional importance (Garonna 2016). This suggests that with climatic warming, temperature limitation of life history timing will be alleviated, with precipitation/moisture limitation increasing in importance. This is represented by the arrow inside the parameter plane in figure 9.1a. Because photoperiodic limitation is unlikely to change, however, if a species or trait is limited by an interaction between photoperiod, temperature, and precipitation that shifts toward greater precipitation limitation and weaker temperature limitation, its position within the parameter plane must remain at a fixed distance from the photoperiod apex.

To the two-dimensional Garonna (2016) parameter plane, we can add an axis representing limitation to, or cuing of, the timing of expression of life history traits by biotic interactions. As was discussed in chapters 6 and 7, interactions among individuals within species, interactions among competing species, and interactions across trophic levels are all influenced by the use of time by the individual in the phenological community. Hence, a logical extension of the Garonna parameter plane is the inclusion of such interactions to derive a three-dimensional parameter space (figure 9.1b). Here again, the location of an individual life history trait of a given species, or the locations of multiple phenophases of a given species, may be positioned within the tetrahedron representing the relative influences of abiotic and biotic constraints on their timing of expression. As well, the relative importance of biotic interactions in limiting or cuing the timing of expression of life history traits may vary interannually as abiotic factors change in importance, or vary spatially across the distribution of a species.

The precise location of life history traits within this parameter space, whether closer to one of its apexes or to the center itself, hence depends upon the importance of one of the principle abiotic limiting factors in relation to the other two and in relation to the importance of biotic interactions. Such variation explains only, however, the factors limiting the timing of expression of life history traits, not why such limitation has evolved. The explanation for the evolution of limitation in each case lies in maximization of the use of time according to the environment in which the trait has evolved and is expressed. The cuing of the timing of expression of life history traits related to development and reproduction by thresholds in photoperiod or solar irradiance, temperature, precipitation or moisture, and biotic interactions all represent convergent evolutionary strategies or solutions to the problem of maximizing the use of relative ecological time to perpetuate genes in cosmological time.

The introduction asked the reader to try imagining how different ecology might be in a universe without dimension—that is, lacking in time and space. Or, perhaps more easily, to try imaging ecology in a universe with space but lacking in time. From chapter 4 onward, this book has explored the consequences for phenological dynamics, species interactions, and biodiversity, of changes in the availability—and use by the individual organism—of absolute cosmological time, recurrent time, and relative ecological time. These consequences clearly scale up to dynamics at higher levels of organization. But does this implicate time as an ecological resource?

Some may argue that this idea is impossible to test empirically, and so impossible to affirm or refute. It may, however, be countered that the experimental manipulation of temperature, snow melt timing, water availability, and CO_2 concentration, as well as analyses of observed changes in such factors in a myriad of phenological studies, many reviewed here, represent such empirical tests. This may be the closest we can come to investigating phenological responses to variation in the availability of time. Additionally, as argued earlier, we may need to reconsider our long-standing notions of resources in ecology. Classically, we consider a resource something that is necessary for growth, survival, and often also reproduction. These descriptors apply to time as convincingly as they apply to any other resource. The difficulty in conceiving of time as a resource, however, relates to the implications of its use by one organism for its use by another organism. Competition, after all, is predicated upon a shared requirement for a limited resource. And this is where our notion of what constitutes a resource in ecology requires refinement, for, in contrast to space and any other resource, the use of time by one organism does not clearly limit its availability for use by another. However, as emphasized earlier, the allocation of time by the individual to a given phenophase or life history transition in its life cycle likely does limit the availability of time for allocation by the same individual to other phenophases or life history transitions. Hence, the strategic allocation of recurrent and relative ecological time *within the individual organism* determines its success or failure in perpetuating its genes through cosmological time, and forms the basis for phenological patterns and dynamics at all higher levels of biological organization. Think of this the next time you notice the arrival of a spring migrant, hear the first call of frogs on a spring evening, or see the first signs of green in a spring forest.

Online Resources of Relevance to Phenology

The following is a list of resources freely accessible online that have direct or indirect relevance to the study of phenology.

1. USA NATIONAL PHENOLOGY NETWORK

www.usanpn.org

USA NPN's mission is to serve "science and society by promoting broad understanding of plant and animal phenology and its relationship with environmental change. The Network is a consortium of individuals and organizations that collect, share, and use phenology data, models, and related information." At the USA NPN website, you will find resources related to the concepts and science of phenology, phenology research communities, education, and data. The Nature's Notebook link provides a portal for sharing phenology data. Nature's Notebook is also available as an app for smartphones (www.usanpn.org/nn /mobile-apps).

2. PROJECT BUDBURST

budburst.org

This online resource provides information on identification of local plant species across regions within the United States, as well as keys to identifying phenophases of local plant species. It also provides resources for teaching the basics of the science of phenology and field applications at all levels of education. Forms are available online for submission of single or ongoing observations of plant phenology. Submissions are provided in standardized format, and data are archived online. Archival data can be accessed through an online portal at the site.

3. U.S. DEPARTMENT OF AGRICULTURE PLANTS DATABASE

http://plants.usda.gov

From the website: "The PLANTS Database provides standardized information about the vascular plants, mosses, liverworts, hornworts, and lichens of the U.S. and its territories."

4. U.S. NATIONAL OCEANIC AND ATMOSPHERIC ADMINISTRATION CLIMATE DATA

www.ncdc.noaa.gov/cdo-web/

5. U.S. NATIONAL SNOW AND ICE DATA CENTER

nsidc.org

6. NORTH AMERICAN LILAC AND HONEYSUCKLE PHENOLOGY DATA

www.nature.com/articles/sdata201538
Published as A. H. Rosemartin, E. G. Denny, J. F. Weltzin, R. L. Marsh, B. E. Wilson, H. Mehdipoor, R. Zurita-Milla, and M. D. Schwartz, 2015, "Lilac and Honeysuckle Phenology Data 1956–2014," *Scientific Data* 2: 150038.

7. NOAA EARTH SYSTEM RESEARCH LABORATORY HIGH-RESOLUTION AIR TEMPERATURE AND PRECIPITATION DATA

www.esrl.noaa.gov/psd/data/gridded/data.UDel_AirT_Precip.html

8. PHENOCAM: AN ECOSYSTEM PHENOLOGY CAMERA NETWORK

phenocam.sr.unh.edu

9. U.S. NATIONAL PARK SERVICE CLEARINGHOUSE FOR ONLINE PHENOLOGY AND EDUCATIONAL RESOURCES

www.nps.gov/seki/learn/education/phenologyresources.htm

10. THE INTERNATIONAL TUNDRA EXPERIMENT

http://ibis.geog.ubc.ca/itex/

11. POLLINATOR LIVE—A DISTANCE LEARNING ADVENTURE

http://pollinatorlive.pwnet.org/teacher/citizen.php

12. *NATURE* KNOWLEDGE PROJECT

www.nature.com/scitable/knowledge

From the website: "The Knowledge Project is a collection of articles on a variety of topics. These articles were developed as a wiki project, independent of Nature Education editors. The Project is no longer active."

There are several modules that are indirectly relevant to phenology, including "Ecosystem Ecology," "Physiological Ecology," "Population Ecology," "Community Ecology," "Global and Regional Ecology," and "Earth's Climate: Past, Present, and Future."

13. UNITED STATES NAVAL OBSERVATORY DURATION OF DAYLIGHT/DARKNESS CALCULATOR

http://aa.usno.navy.mil/data/docs/Dur_OneYear.php

Allows the user to calculate the total length of time that the Sun is at least partially above the horizon for any date at any location on the Earth.

14. APPLES—ARCTIC PLANT PHENOLOGY: LEARNING THROUGH ENGAGED SCIENCE

www.applesproject.org

U.S. National Science Foundation–supported project engaging K–12 educators in plant phenology research, teaching, and outreach related to climate change. Offers the opportunity for participating educators and students to share their own plant phenology data and educational modules, as well as access to long-term data on plant phenology from the author's research near Kangerlussuaq, Greenland.

Sources Used in the
Meta-analysis in Chapter 2

Source	Taxa
Source	*Taxa*
Adamik and Kral (2008)	birds, mammals
Adamik and Pietruszkova (2008)	birds
Ahas (1999)	birds, fish
Beebee (1995)	amphibians
Beebee (2007)	amphibians
Bergant et al. (2002)	plants
Bergmann (1999)	birds
Bertram et al. (2001)	birds
Blaustein et al. (2001)	amphibians
Both and Visser (2001)	birds
Bradley et al. (1999)	birds, plants
Brown et al. (1999)	birds
Carroll et al. (2009)	amphibians
Cayan et al. (2001)	plants
Chmielewski and Rotzer (2001)	plants
Chuine et al. (2000)	plants
Cook et al. (2008)	amphibians, plants, invertebrates
Crick and Sparks (1999)	birds
Crick et al. (1996)	birds
Croxton et al. (2006)	birds
de Beurs and Henebry (2005)	plants
Defila and Clot (2001)	plants
Diamond et al. (2011)	invertebrates

Source	Taxa
Dunn and Winkler (1999)	birds
Elliott (1996)	invertebrates
Emberlin et al. (1999)	plants
Fleming and Tatchell (1995)	invertebrates
Gordo et al. (2005)	birds
Høye et al. (2007)	birds, invertebrates, plants
Inouye et al. (2000)	birds, mammals
Jarvinen (1989)	birds
Jeong et al. (2009)	plants
Julien and Sobrino (2009)	plants
Kusano and Inoue (2008)	amphibians
Lambert et al. (2010)	plants
Lane et al. (2012)	mammals
Ludwichowski (1997)	birds
Ma and Zhou (2012)	plants
Macinnes et al. (1990)	birds
Mackas et al. (1998)	invertebrates
McCleery and Perrins (1998)	birds
Menzel and Fabian (1999)	plants
Menzel et al. (2001)	plants
Mickelson et al. (1992)	birds
Myneni et al. (1997)	plants
Osborne et al. (2000)	plants
Ozgul et al. (2010)	mammals
Penuelas et al. (2002)	birds, invertebrates, plants
Piao et al. (2006)	plants
Pollard (1991)	invertebrates
Post et al. (2016)	plants
Reading (1998)	amphibians
Reale et al. (2003)	mammals
Roetzer et al. (2000)	plants
Roy and Sparks (2000)	invertebrates
Schiegg et al. (2002)	birds

Source	Taxa
Source	*Taxa*
Schwartz (1998)	plants
Schwartz and Caprio (2003)	plants
Schwartz and Reiter (2000)	plants
Slater (1999)	birds
Spano et al. (1999)	plants
Sparks (1999)	birds
Sparks and Braslavska (2001)	birds
Sparks et al. (2007)	birds
Stöckli and Vidale (2004)	plants
Thompson and Clark (2008)	plants
Thompson et al. (1986)	birds
Tryjanowski et al. (2003)	amphibians
Tucker et al. (2001)	plants
Vegvari et al. (2010)	birds
Visser et al. (1998)	birds, invertebrates
Winkel and Hudde (1996)	birds
Zalakevicius et al. (2006)	birds

References

Adamik, P., and M. Kral. 2008. "Climate- and Resource-driven Long-term Changes in Dormice Populations Negatively Affect Hole-nesting Songbirds." *Journal of Zoology* 275: 209–215.

Adamik, P., and J. Pietruszkova. 2008. "Advances in Spring but Variable Autumnal Trends in Timing of Inland Wader Migration." *Acta Ornithologica* 43: 119–128.

Adler, P. B., R. Salguero-Gomez, A. Compagnoni, J. S. Hsu, J. Ray-Mukherjee, C. Mbeau-Ache, and M. Franco. 2014. "Functional Traits Explain Variation in Plant Life History Strategies." *Proceedings of the National Academy of Sciences of the United States of America* 111: 740–745.

Ahas, R. 1999. "Long-term Phyto-, Ornitho- and Ichthyophenological Time-series Analyses in Estonia." *International Journal of Biometeorology* 42: 119–123.

Aizen, M. A., and A. E. Rovere. 2010. "Reproductive Interactions Mediated by Flowering Overlap in a Temperate Hummingbird-Plant Assemblage." *Oikos* 119: 696–706.

Albon, S. D., and R. Langvatn. 1992. "Plant Phenology and the Benefits of Migration in a Temperate Ungulate." *Oikos* 65: 502–513.

Allstadt, A. J., S. J. Vavrus, P. J. Heglund, A. M. Pidgeon, W. E. Thogmartin, and V. C. Radeloff. 2015. "Spring Plant Phenology and False Springs in the Conterminous US during the 21st Century." *Environmental Research Letters* 10: 104008.

Altermatt, F. 2010. "Climatic Warming Increases Voltinism in European Butterflies and Moths." *Proceedings of the Royal Society B—Biological Sciences* 277: 1281–1287.

———. 2012. "Temperature-related Shifts in Butterfly Phenology Depend on the Habitat." *Global Change Biology* 18: 2429–2438.

Archetti, M., A. D. Richardson, J. O'Keefe, and N. Delpierre. 2013. "Predicting Climate Change Impacts on the Amount and Duration of Autumn Colors in a New England Forest." *PLoS One* 8(3): e57373.

Aronson, B. D., D. Bellpedersen, G. D. Block, N.P.A. Bos, J. C. Dunlap, A. Eskin, N. Y. Garceau, M. E. Geusz, K. A. Johnson, S.B.S. Khalsa, G. C. Kostervanhoffen, C. Koumenis, T. M. Lee, J. Lesauter, K. M. Lindgren, Q. Y. Liu, J. J. Loros, S. H. Michel, M. Mirmiran, R. Y. Moore, N. F. Ruby, R. Silver, F. W. Turek, M. Zatz, and I. Zucker. 1993. "Circadian-Rhythms." *Brain Research Reviews* 18: 315–333.

Aschoff, J. 1960. "Exogenous and Endogenous Components in Circadian Rhythms." *Cold Spring Harbor Symposia on Quantitative Biology* 25: 11–28.

———. 1966. "Circadian Activity Pattern with 2 Peaks." *Ecology* 47: 657–662.

Avery, M., and E. Post. 2013. "Record of a *Zoophthora* Sp. (Entomophthoromycota: *Entomophthorales*) Pathogen of the Irruptive Noctuid Moth *Eurois occulta* (*Lepidoptera*) in West Greenland." *Journal of Invertebrate Pathology* 114: 292–294.

Badeck, F. W., A. Bondeau, K. Bottcher, D. Doktor, W. Lucht, J. Schaber, and S. Sitch. 2004. "Responses of Spring Phenology to Climate Change. *New Phytologist* 162: 295–309.

Bardon, A. 2013. *A Brief History of the Philosophy of Time*. Oxford, UK: Oxford University Press.

Bartomeus, I., J. S. Ascher, J. Gibbs, B. N. Danforth, D. L. Wagner, S. M. Hedtke, and R. Winfree. 2013. "Historical Changes in Northeastern US Bee Pollinators Related to Shared Ecological Traits." *Proceedings of the National Academy of Sciences of the United States of America* 110: 4656–4660.

Bartomeus, I., J. S. Ascher, D. Wagner, B. N. Danforth, S. Colla, S. Kornbluth, and R. Winfree. 2011. "Climate-associated Phenological Advances in Bee Pollinators and Bee-pollinated Plants." *Proceedings of the National Academy of Sciences of the United States of America* 108: 20645–20649.

Beaumont, L. J., I.A.W. McAllan, and L. Hughes. 2006. "A Matter of Timing: Changes in the First Date of Arrival and Last Date of Departure of Australian Migratory Birds." *Global Change Biology* 12: 1339–1354.

Beebee, T.J.C. 1995. "Amphibian Breeding and Climate." *Nature* 374: 219–220.

———. 2007. "Thirty Years of Garden Ponds. *Herpetological Bulletin* 99: 23–28.

Begon, M., C. R. Townsend, and J. L. Harper. 2006. *Ecology: From Individuals to Ecosystems*. London: Wiley-Blackwell.

Bentz, B. J., and J. A. Powell. 2014. "Mountain Pine Beetle Seasonal Timing and Constraints to Bivoltinism (A Comment on Mitton and Ferrenberg, 'Mountain Pine Beetle Develops an Unprecedented Summer Generation in Response to Climate Warming')." *American Naturalist* 184: 787–796.

Bergant, K., L. Kajfez-Bogataj, and Z. Crepinsek. 2002. "Statistical Downscaling of General-Circulation-Model-Simulated Average Monthly Air Temperature to the Beginning of Flowering of the Dandelion (*Taraxacum officinale*) in Slovenia." *International Journal of Biometeorology* 46: 22–32.

Bergmann, F. 1999. "Long-term Increase in Numbers of Early-fledged Reed Warblers (*Acrocephalus scirpaceus*) at Lake Constance (Southern Germany)." *Journal fur Ornithologie* 140: 81–86.

Bertram, D. F., D. L. Mackas, and S. M. McKinnell. 2001. "The Seasonal Cycle Revisited: Interannual Variation and Ecosystem Consequences." *Progress in Oceanography* 49: 283–307.

Biesmeijer, J. C., S.P.M. Roberts, M. Reemer, R. Ohlemueller, M. Edwards, T. Peeters, A. P. Schaffers, S. G. Potts, R. Kleukers, C. D. Thomas, J. Settele, and W. E. Kunin. 2006. "Parallel Declines in Pollinators and Insect-pollinated Plants in Britain and the Netherlands." *Science* 313: 351–354.

Billings, W. D., K. M. Peterson, G. R. Shaver, and A. W. Trent. 1977. "Root Growth, Respiration, and Carbon-dioxide Evolution in an Arctic Tundra Soil." *Arctic and Alpine Research* 9: 129–137.

Bjorkman, A. D., S. C. Elmendorf, A. L. Beamish, M. Vellend, and G.H.R. Henry. 2015. "Contrasting Effects of Warming and Increased Snowfall on Arctic Tundra Plant Phenology over the Past Two Decades." *Global Change Biology* 21: 4651–4661.

Bjørnstad, O. N., B. F. Finkenstadt, and B. T. Grenfell. 2002. "Dynamics of Measles Epidemics: Estimating Scaling of Transmission Rates Using a Time Series SIR Model." *Ecological Monographs* 72: 169–184.

Bjørnstad, O. N., and B. T. Grenfell. 2001. "Noisy Clockwork: Time Series Analysis of Population Fluctuations in Animals." *Science* 293: 638–643.

Blaustein, A. R., L. K. Belden, D. H. Olson, D. M. Green, T. L. Root, and J. M. Kie-secker. 2001. "Amphibian Breeding and Climate Change." *Conservation Biology* 15: 1804–1809.

Bluethgen, N., and A.-M. Klein. 2011. "Functional Complementarity and Specialisation: The Role of Biodiversity in Plant-Pollinator Interactions." *Basic and Applied Ecology* 12: 282–291.

Bock, A., T. H. Sparks, N. Estrella, N. Jee, A. Casebow, C. Schunk, M. Leuchner, and A. Menzel. 2014. "Changes in First Flowering Dates and Flowering Duration of 232 Plant Species on the Island of Guernsey." *Global Change Biology* 20: 3508–3519.

Bokhorst, S., J. W. Bjerke, L. E. Street, T. V. Callaghan, and G. K. Phoenix. 2011. "Impacts of Multiple Extreme Winter Warming Events on Sub-Arctic Heathland: Phenology, Reproduction, Growth, and CO_2 Flux Responses." *Global Change Biology* 17: 2817–2830.

Both, C., R. G. Bijlsma, and M. E. Visser. 2005. "Climatic Effects on Timing of Spring Migration and Breeding in a Long-distance Migrant, the Pied Flycatcher *Ficedula Hypoleuca*." *Journal of Avian Biology* 36: 368–373.

Both, C., S. Bouwhuis, C. M. Lessells, M. E. Visser. 2006. "Climate Change and Population Declines in a Long-distance Migratory Bird." *Nature* 441: 81–83.

Both, C., M. Van Asch, R. G. Bijlsma, A. B. Van Den Burg, and M. E. Visser. 2009. "Climate Change and Unequal Phenological Changes across Four Trophic Levels: Constraints or Adaptations?" *Journal of Animal Ecology* 78: 73–83.

Both, C., and M. E. Visser. 2001. "Adjustment to Climate Change Is Constrained by Arrival Date in a Long-distance Migrant Bird." *Nature* 411: 296–298.

———. 2005. "The Effect of Climate Change on the Correlation between Avian Life-history Traits." *Global Change Biology* 11: 1606–1613.

Bøving, P. S., and E. Post. 1997. "Vigilance and Foraging Behaviour of Female Caribou in Relation to Predation Risk." *Rangifer* 17: 55–63.

Bradley, N. L., A. C. Leopold, J. Ross, and W. Huffaker. 1999. "Phenological Changes Reflect Climate Change in Wisconsin." *Proceedings of the National Academy of Sciences of the United States of America* 96: 9701–9704.

Brando, P., D. Ray, D. Nepstad, G. Cardinot, L. M. Curran, and R. Oliveira. 2006. "Effects of Partial Throughfall Exclusion on the Phenology of *Coussarea racemosa* (Rubiaceae) in an East-Central Amazon Rainforest." *Oecologia* 150: 181–189.

Brearley, F. Q., J. Proctor, Suriantata, L. Nagy, G. Dalrymple, and B. C. Voysey. 2007. "Reproductive Phenology over a 10-year Period in a Lowland Evergreen Rain Forest of Central Borneo." *Journal of Ecology* 95: 828–839.

Breeman, A. M., E.J.S. Meulenhoff, and M. D. Guiry. 1988. "Life-history Regulation and Phenology of the Red Alga *Bonnemaisonia Hamifera*." *Helgolander meeresuntersuchungen* 42: 535–551.

Brittain, C. A., M. Vighi, R. Bommarco, J. Settele, and S. G. Potts. 2010. "Impacts of a Pesticide on Pollinator Species Richness at Different Spatial Scales." *Basic and Applied Ecology* 11: 106–115.

Brodie, J., E. Post, and W. Laurance. 2010. "How to Conserve the Tropics as They Warm." *Nature* 468: 634–634.

———. 2012. "Climate Change and Tropical Biodiversity: A New Focus." *Trends in Ecology and Evolution* 27: 145–150.

Brown, J. H., J. F. Gillooly, A. P. Allen, V. M. Savage, and G. B. West. 2004. "Toward a Metabolic Theory of Ecology." *Ecology* 85: 1771–1789.

Brown, J. H., and E. J. Heske. 1990. "Temporal Changes in a Chihuahuan Desert Rodent Community." *Oikos* 59: 290–302.

Brown, J. H., and J. C. Munger. 1985. "Experimental Manipulation of a Desert Rodent Community—Food Addition and Species Removal." *Ecology* 66: 1545–1563.

Brown, J. L., S. H. Li, and N. Bhagabati. 1999. "Long-term Trend toward Earlier Breeding in an American Bird: A Response to Global Warming?" *Proceedings of the National Academy of Sciences of the United States of America* 96: 5565–5569.

Buitenwerf, R., L. Rose, and S. I. Higgins. 2015. "Three Decades of Multi-dimensional Change in Global Leaf Phenology." *Nature Climate Change* 5: 364–368.

Bullock, S. H., and J. A. Solismagallanes. 1990. "Phenology of Canopy Trees of a Tropical Deciduous Forest in Mexico." *Biotropica* 22: 22–35.

Bulluck, L., S. Huber, C. Viverette, and C. Blem. 2013. "Age-specific Responses to Spring Temperature in a Migratory Songbird: Older Females Attempt More Broods in Warmer Springs." *Ecology and Evolution* 3: 3298–3306.

Burkle, L. A., and R. Alarcon. 2011. "The Future of Plant-Pollinator Diversity: Understanding Interaction Networks across Time, Space, and Global Change." *American Journal of Botany* 98: 528–538.

Burkle, L. A., J. C. Marlin, and T. M. Knight. 2013. "Plant-Pollinator Interactions over 120 Years: Loss of Species, Co-occurrence, and Function." *Science* 339: 1611–1615.

Butt, N., L. Seabrook, M. Maron, B. S. Law, T. P. Dawson, J. Syktus, and C. A. McAlpine. 2015. "Cascading Effects of Climate Extremes on Vertebrate Fauna through Changes to Low-latitude Tree Flowering and Fruiting Phenology." *Global Change Biology* 21: 3267–3277.

Callender, C. 2011. "Introduction—The Direction of Time." In *The Oxford Handbook of Philosophy of Time*, ed. C. Callender. Oxford, UK: Oxford University Press, 1–12.

Caprio, J. M. 1957. "Phenology of Lilac Bloom in Montana." *Science* 126: 1344–1345.

CaraDonna, P. J., A. M. Iler, and D. W. Inouye. 2014. "Shifts in Flowering Phenology Reshape a Subalpine Plant Community." *Proceedings of the National Academy of Sciences of the United States of America* 111: 4916–4921.

Carpenter, S. R., J. J. Cole, J. R. Hodgson, J. F. Kitchell, M. L. Pace, D. Bade, K. L. Cottingham, T. E. Essington, J. N. Houser, and D. E. Schindler. 2001. "Trophic Cascades, Nutrients, and Lake Productivity: Whole-lake Experiments." *Ecological Monographs* 71: 163–186.

Carpenter, S. R., J. J. Cole, M. L. Pace, R. Batt, W. A. Brock, T. Cline, J. Coloso, J. R. Hodgson, J. F. Kitchell, D. A. Seekell, L. Smith, and B. Weidel. 2011. "Early Warnings of Regime Shifts. A Whole-Ecosystem Experiment." *Science* 332: 1079–1082.

Carpenter, S. R., and J. F. Kitchell. 1988. "Consumer Control of Lake Productivity." *Bioscience* 38: 764–769.

Carpenter, S. R., J. F. Kitchell, and J. R. Hodgson. 1985. "Cascading Trophic Interactions and Lake Productivity." *Bioscience* 35: 634–639.

Carroll, E. A., T. H. Sparks, N. Collinson, and T.J.C. Beebee. 2009. "Influence of Temperature on the Spatial Distribution of First Spawning Dates of the Common Frog (*Rana Temporaria*) in the UK." *Global Change Biology* 15: 467–473.

Carvalho, D. M., S. J. Presley, and G.M.M. Santos. 2014. "Niche Overlap and Network Specialization of Flower-visiting Bees in an Agricultural System." *Neotropical Entomology* 43: 489–499.

Cayan, D. R., S. A. Kammerdiener, M. D. Dettinger, J. M. Caprio, and D. H. Peterson. 2001. "Changes in the Onset of Spring in the Western United States." *Bulletin of the American Meteorological Society* 82: 399–415.

Chapman, C. A., L. J. Chapman, T. T. Struhsaker, A. E. Zanne, C. J. Clark, and J. R. Poulsen. 2005. "A Long-term Evaluation of Fruiting Phenology: Importance of Climate Change." *Journal of Tropical Ecology* 21: 31–45.

Chesson, P. 2000. "Mechanisms of Maintenance of Species Diversity." *Annual Review of Ecology and Systematics* 31: 343–366.

Chesson, P., and N. Huntly. 1989. "Short-term Instabilities and Long-term Community Dynamics." *Trends in Ecology and Evolution* 4: 293–298.

Chmielewski, F. M., and T. Rotzer. 2001. "Response of Tree Phenology to Climate Change across Europe." *Agricultural and Forest Meteorology* 108: 101–112.

Chuine, I., G. Cambon, and P. Comtois. 2000. "Scaling Phenology from the Local to the Regional Level: Advances from Species-specific Phenological Models." *Global Change Biology* 6: 943–952.

Clausen, K. K., and P. Clausen. 2013. "Earlier Arctic Springs Cause Phenological Mismatch in Long-distance Migrants." *Oecologia* 173: 1101–1112.

Connell, J. H. 1961. "The Influence of Interspecific Competition and Other Factors on the Distribution of the Barnacle *Chthamalus stellatus*." *Ecology* 42: 710–723.

Cook, B. I., E. R. Cook, P. C. Huth, J. E. Thompson, A. Forster, and D. Smiley. 2008. "A Cross-taxa Phenological Dataset from Mohonk Lake, NY, and Its Relationship to Climate." *International Journal of Climatology* 28: 1369–1383.

Corlett, R. T., and J. V. LaFrankie. 1998. "Potential Impacts of Climate Change on Tropical Asian Forests through an Influence on Phenology." *Climatic Change* 39: 439–453.

Crick, H., C. Dudley, and D. Glue. 1996. "Nesting in 1993." *Bird Populations* 3: 165–169.

Crick, H.Q.P., and T. H. Sparks. 1999. "Climate Change Related to Egg-laying Trends." *Nature* 399: 423–424.

Crimmins, T. M., C. D. Bertelsen, and M. A. Crimmins. 2014. "Within-season Flowering Interruptions Are Common in the Water-limited Sky Islands." *International Journal of Biometeorology* 58: 419–426.

Crimmins, T. M., M. A. Crimmins, and C. D. Bertelsen. 2013. "Spring and Summer Patterns in Flowering Onset, Duration, and Constancy across a Water-limited Gradient." *American Journal of Botany* 100: 1137–1147.

Croxton, P. J., T. H. Sparks, M. Cade, and R. G. Loxton. 2006. "Trends and Temperature Effects in the Arrival of Spring Migrants in Portland (United Kingdom) 1959–2005." *Acta Ornithologica* 41: 103–111.

Curran, L. M., and M. Leighton. 2000. "Vertebrate Responses to Spatiotemporal Variation in Seed Production of Mast-fruiting Dipterocarpaceae." *Ecological Monographs* 70: 101–128.

Dante, S. K., B. S. Schamp, and L. W. Aarssen. 2013. "Evidence of Deterministic Assembly According to Flowering Time in an Old-field Plant Community." *Functional Ecology* 27: 555–564.

de Beurs, K. M., and G. M. Henebry. 2005. "Land Surface Phenology and Temperature Variation in the International Geosphere-Biosphere Program High-latitude Transects." *Global Change Biology* 11: 779–790.

Defila, C., and B. Clot. 2001. "Phytophenological Trends in Switzerland." *International Journal of Biometeorology* 45: 203–207.

Diamond, S. E., H. Cayton, T. Wepprich, C. N. Jenkins, R. R. Dunn, N. M. Haddad, and L. Ries. 2014. "Unexpected Phenological Responses of Butterflies to the Interaction of Urbanization and Geographic Temperature." *Ecology* 95: 2613–2621.

Diamond, S. E., A. M. Frame, R. A. Martin, and L. B. Buckley. 2011. "Species' Traits Predict Phenological Responses to Climate Change in Butterflies." *Ecology* 92: 1005–1012.

Diez, J. M., I. Ibanez, A. J. Miller-Rushing, S. J. Mazer, T. M. Crimmins, M. A. Crimmins, C. D. Bertelsen, and D. W. Inouye. 2012. "Forecasting Phenology: from Species Variability to Community Patterns." *Ecology Letters* 15: 545–553.

Doi, H. 2008. "Delayed Phenological Timing of Dragonfly Emergence in Japan over Five Decades." *Biology Letters* 4: 388–391.

Dorji, T., O. Totland, S. R. Moe, K. A. Hopping, J. B. Pan, and J. A. Klein. 2013. "Plant Functional Traits Mediate Reproductive Phenology and Success in Response to Experimental Warming and Snow Addition in Tibet." *Global Change Biology* 19: 459–472.

Dunham, A. E., E. M. Erhart, and P. C. Wright. 2011. "Global Climate Cycles and Cyclones: Consequences for Rainfall Patterns and Lemur Reproduction in Southeastern Madagascar." *Global Change Biology* 17: 219–227.

Dunn, P. O., and D. W. Winkler. 1999. "Climate Change Has Affected the Breeding Date of Tree Swallows throughout North America." *Proceedings of the Royal Society of London Series B—Biological Sciences* 266: 2487–2490.

Dutka, A., A. McNulty, and S. M. Williamson. 2015. "A New Threat to Bees? Entomopathogenic Nematodes Used in Biological Pest Control Cause Rapid Mortality in Bombus Terrestris." *Peerj* 3: e1413.

Egevang, C., I. J. Stenhouse, R. A. Phillips, A. Petersen, J. W. Fox, and J.R.D. Silk. 2010. "Tracking of Arctic Terns *Sterna paradisaea* Reveals Longest Animal Migration." *Proceedings of the National Academy of Sciences of the United States of America* 107: 2078–2081.

Ehrlen, J. 2015. "Selection on Flowering Time in a Life-cycle Context." *Oikos* 124: 92–101.

Elith, J., and J. R. Leathwick. 2009. "Species Distribution Models: Ecological Explanation and Prediction across Space and Time." *Annual Review of Ecology Evolution and Systematics 40: 677-697.*

Ellison, A. M., N. J. Gotelli, J. S. Brewer, D. L. Cochran-Stafira, J. M. Kneitel, T. E. Miller, A. C. Worley, and R. Zamora. 2003. "The Evolutionary Ecology of Carnivorous Plants." *Advances in Ecological Research* 33: 1-74.

Elliott, J. M. 1996. "Temperature-related Fluctuations in the Timing of Emergence and Pupation of Windermere Alder-flies over 30 Years." *Ecological Entomology* 21: 241–247.

Ellwood, E. R., S. A. Temple, R. B. Primack, N. L. Bradley, and C. C. Davis. 2013. "Record-Breaking Early Flowering in the Eastern United States." *PLoS One* 8(1): e53788.

Elton, C. S. 1958. *The Ecology of Invasions by Animals and Plants*. London: Methuen.

Elton, C. S., and R. S. Miller. 1954. "The Ecological Survey of Animal Communities—With a Practical System of Classifying Habitats by Structural Characters." *Journal of Ecology* 42: 460–496.

Emberlin, J., J. Mullins, J. Corden, S. Jones, W. Millington, M. Brooke, and M. Savage. 1999. "Regional Variations in Grass Pollen Seasons in the UK, Long-term Trends and Forecast Models." *Clinical and Experimental Allergy* 29: 347–356.

Engelhardt, M. J., and R. C. Anderson. 2011. "Phenological Niche Separation from Native Species Increases Reproductive Success of an Invasive Species: *Alliaria petiolata* (*Brassicaceae*)—Garlic Mustard." *Journal of the Torrey Botanical Society* 138: 418–433.

Enquist, B. J., E. P. Economo, T. E. Huxman, A. P. Allen, D. D. Ignace, and J. F. Gillooly. 2003. "Scaling Metabolism from Organisms to Ecosystems." *Nature* 423: 639–642.

Ernakovich, J. G., K. A. Hopping, A. B. Berdanier, R. T. Simpson, E. J. Kachergis, H. Steltzer, and M. D. Wallenstein. 2014. "Predicted Responses of Arctic and Alpine Ecosystems to Altered Seasonality under Climate Change." *Global Change Biology* 20: 3256–3269.

Fabina, N. S., K. C. Abbott, and R. T. Gilman. 2010. "Sensitivity of Plant-Pollinator-Herbivore Communities to Changes in Phenology." *Ecological Modelling* 221: 453–458.

Fargione, J., and D. Tilman. 2005. "Niche Differences in Phenology and Rooting Depth Promote Coexistence with a Dominant C-4 Bunchgrass." *Oecologia* 143: 598–606.

Fisher, R. A. 1930. *The Genetical Theory of Natural Selection*. New York: Dover.

Fleming, R. A., and G. M. Tatchell. 1995. "Shifts in the Flight Periods of British Aphids: A Response to Climate Warming?" In *Insects in a Changing Environment*, ed. R. Harrington and N. E. Stork. San Diego, CA: Academic Press, 505–508.

Fleming, T. H., R. Breitwisch, and G. H. Whitesides. 1987. "Patterns of Tropical Vertebrate Frugivore Diversity." *Annual Review of Ecology and Systematics* 18: 91–109.

Fleming, T. H., and B. L. Partridge. 1984. "On the Analysis of Phenological Overlap." *Oecologia* 62: 344–350.

Forchhammer, M. C., T. R. Christensen, B. U. Hansen, M. T. Tamstorf, J. Hinkler, N. M. Schmidt, T. T. Høye, M. Rasch, H. Meltofte, B. Elberling, and E. Post. 2008a. "Zackenberg in a Circumpolar Context." *Advances in Ecological Research* 40: 499–544.

Forchhammer, M. C., and E. Post. 2004. "Using Large-scale Climate Indices in Climate Change Ecology Studies." *Population Ecology* 46: 1–12.

Forchhammer, M. C., E. Post, T.B.G. Berg, T. T. Hoye, and N. M. Schmidt. 2005. "Local-scale and Short-term Herbivore-Plant Spatial Dynamics Reflect Influences of Large-scale Climate." *Ecology* 86: 2644–2651.

Forchhammer, M. C., E. Post, and N. C. Stenseth. 1998. "Breeding Phenology and Climate." *Nature* 391: 29–30.

———. 2002. "North Atlantic Oscillation Timing of Long- and Short-distance Migration." *Journal of Animal Ecology* 71: 1002–1014.

Forchhammer, M. C., N. M. Schmidt, T. T. Høye, T. B. Berg, D. K. Hendrichsen, and E. Post. 2008b. "Population Dynamical Responses to Climate Change." *Advances in Ecological Research* 40: 391–419.

Forister, M. L., and A. M. Shapiro. 2003. "Climatic Trends and Advancing Spring Flight of Butterflies in Lowland California." *Global Change Biology* 9: 1130–1135.

Forister, M. L., and A. M. Shapiro. In Press. "Spring Flight of California Central Valley Butterflies (updated)." In *Indicators of Climate Change in California*, ed. T. Kadir, L. Mazur, C. Milanes, and K. Randles. Sacramento, CA: California Environmental Protection Agency, Office of Environmental Health Hazard Assessment.

Forrest, J., and A. J. Miller-Rushing. 2010. "Toward a Synthetic Understanding of the Role of Phenology in Ecology and Evolution." *Philosophical Transactions of the Royal Society B—Biological Sciences* 365: 3101–3112.

Forrest, J.R.K. 2015. "Plant-Pollinator Interactions and Phenological Change: What Can We Learn about Climate Impacts from Experiments and Observations?" *Oikos* 124: 4–13.

———. 2016. "Complex Responses of Insect Phenology to Climate Change." *Current Opinion in Insect Science* 17: 49–54.

Forrest, J.R.K., and J. D. Thomson. 2011. "An Examination of Synchrony between Insect Emergence and Flowering in Rocky Mountain Meadows." *Ecological Monographs* 81: 469–491.

Framstad, E., N. C. Stenseth, O. N. Bjørnstad, and W. Falck. 1997. "Limit Cycles in Norwegian Lemmings: Tensions between Phase-dependence and Density-dependence." *Proceedings of the Royal Society of London, Series B* 264: 31–38.

Frankie, G. W., H. G. Baker, and P. A. Opler. 1974. "Comparative Phenological Studies of Trees in Tropical Wet and Dry Forests in Lowlands of Costa Rica." *Journal of Ecology* 62: 881–919.

Fredriksson, G. M., L. S. Danielsen, and J. E. Swenson. 2007. "Impacts of El Nino Related Drought and Forest Fires on Sun Bear Fruit Resources in Lowland Dipterocarp Forest of East Borneo." *Biodiversity and Conservation* 16: 1823–1838.

Fredriksson, G. M., S. A. Wich, and TRISNO. 2006. "Frugivory in Sun Bears (*Helarctos malayanus*) Is Linked to El Nino Related Fluctuations in Fruiting Phenology, East Kalimantan, Indonesia." *Biological Journal of the Linnean Society* 89: 489–508.

Gadgil, M., and W. H. Bossert. 1970. "Life Historical Consequences of Natural Selection." *American Naturalist* 104: 1–24.

Galen, C., and M. L. Stanton. 1995. "Responses of Snowbed Plant Species to Changes in Growing-season Length." *Ecology* 76: 1546–1557.

Gallai, N., J.-M. Salles, J. Settele, and B. E. Vaissiere. 2009. "Economic Valuation of the Vulnerability of World Agriculture Confronted with Pollinator Decline." *Ecological Economics* 68: 810–821.

Garonna, I. 2016. "Global Vegetation Trends." Ph.D. Dissertation, University of Zurich, Switzerland.

Garonna, I., R. De Jong, A.J.W. De Wit, C. A. Mucher, B. Schmid, and M. E. Schaepman. 2014. "Strong Contribution of Autumn Phenology to Changes in Satellite-derived Growing Season Length Estimates across Europe (1982–2011)." *Global Change Biology* 20: 3457–3470.

Garonna, I., R. De Jong, and M. E. Schaepman. 2016. "Variability and Evolution of Global Land Surface Phenology over the Past Three Decades (1982–2012)." *Global Change Biology* 22(4): 1456–1468.

Gaston, K. J. 2000. "Global Patterns in Biodiversity." *Nature* 405: 220–227.

Gilman, R. T., N. S. Fabina, K. C. Abbott, and N. E. Rafferty. 2012. "Evolution of Plant-Pollinator Mutualisms in Response to Climate Change." *Evolutionary Applications* 5: 2–16.

Goff, L. J., and K. Cole. 1976. "Biology of *Harveyella-mirabilis* (*Cryptonemiales-Rhodophyceae*). 4. Life-History and Phenology." *Canadian Journal of Botany—Revue Canadienne de Botanique* 54: 281–292.

Gordo, O., L. Brotons, X. Ferrer, and P. Comas. 2005. "Do Changes in Climate Patterns in Wintering Areas Affect the Timing of the Spring Arrival of Trans-Saharan Migrant Birds?" *Global Change Biology* 11: 12–21.

Gotelli, N. J., M. J. Anderson, H. T. Arita, A. Chao, R. K. Colwell, S. R. Connolly, D. J. Currie, R. R. Dunn, G. R. Graves, J. L. Green, J. A. Grytnes, Y. H. Jiang, W. Jetz, S. K. Lyons, C. M. McCain, A. E. Magurran, C. Rahbek, T. Rangel, J. Soberon, C. O. Webb, and M. R. Willig. 2009. "Patterns and Causes of Species Richness: A General Simulation Model for Macroecology." *Ecology Letters* 12: 873–886.

Gotelli, N. J., and D. J. McCabe. 2002. "Species Co-occurrence: A Meta-analysis of J. M. Diamond's Assembly Rules Model." *Ecology* 83: 2091–2096.

Grenfell, B. T., O. N. Bjørnstad, and J. Kappey. 2001. "Travelling Waves and Spatial Hierarchies in Measles Epidemics." *Nature* 414: 716–723.

Grinnell, J. 1917. "The Niche-relationships of the California Thrasher." *Auk* 34: 427–433.

Gunarathne, R.M.U.K., and G.A.D. Perera. 2014. "Climatic Factors Responsible For Triggering Phenological Events in *Manilkara hexandra* (*Roxb.*) *Dubard.*, a Canopy Tree in Tropical Semi-deciduous Forest of Sri Lanka." *Tropical Ecology* 55: 63–73.

Harris, G. A. 1977. "Root Phenology as a Factor of Competition among Grass Seedlings." *Journal of Range Management* 30: 172–177.

Hart, R., J. Salick, S. Ranjitkar, and J. C. Xu. 2014. "Herbarium Specimens Show Contrasting Phenological Responses to Himalayan Climate." *Proceedings of the National Academy of Sciences of the United States of America* 111: 10615–10619.

Harte, J., and R. Shaw. 1995. "Shifting Dominance within a Montane Vegetation Community—Results of a Climate-warming Experiment." *Science* 267: 876–880.

Hastie, T. J., and R. J. Tibshirani. 1999. *Generalized Additive Models*. London: Chapman and Hall/CRC.

Hawking, S. 1988. *A Brief History of Time*. London: Bantam Books.

———. 1996a. "Classical Theory." In *The Nature of Space and Time*, ed. S. Hawking and R. Penrose. Princeton, NJ: Princeton University Press, 3–26.

———. 1996b. "Quantum Black Holes." In *The Nature of Space and Time*, ed. S. Hawking and R. Penrose. Princeton, NJ: Princeton University Press, 37–60.

———. 1996c. "Quantum Cosmology." In *The Nature of Space and Time*, ed. S. Hawking and R. Penrose. Princeton, NJ: Princeton University Press, 75–104.

Hawking, S., and R. Penrose, eds. 1996. *The Nature of Space and Time*. Princeton, NJ: Princeton University Press.

Hawking, S. W. 1969. "Existence of Cosmic Time Functions." *Proceedings of the Royal Society of London Series A—Mathematical and Physical Sciences* 308: 433–435.

———. 1985. "Arrow of Time in Cosmology." *Physical Review D* 32: 2489–2495.

———. 1990. "Events in Time." *Parabola—Myth Tradition and the Search for Meaning* 15: 73–76.

Hebblewhite, M., E. Merrill, and G. McDermid. 2008. "A Multi-scale Test of the Forage Maturation Hypothesis in a Partially Migratory Ungulate Population." *Ecological Monographs* 78: 141–166.

Hegazy, A. K., A. A. Alatar, J. Lovett-Doust, and H. A. El-Adawy. 2012. "Spatial and Temporal Plant Phenological Niche Differentiation in the Wadi Degla Desert Ecosystem (Egypt)." *Acta Botanica Croatica* 71: 261–277.

Hegland, S. J., A. Nielsen, A. Lazaro, A.-L. Bjerknes, and O. Totland. 2009. "How Does Climate Warming Affect Plant-Pollinator Interactions?" *Ecology Letters* 12: 184–195.

Heithaus, E. R. 1979. "Flower-feeding Specialization in Wild Bee and Wasp Communities in Seasonal Neotropical Habitats." *Oecologia* 42: 179–194.

Hindell, M. A., C.J.A. Bradshaw, B. W. Brook, D. A. Fordham, K. Kerry, C. Hull, and C. R. McMahon. 2012. "Long-term Breeding Phenology Shift in Royal Penguins." *Ecology and Evolution* 2: 1563–1571.

Hocking, B. 1968. "Insect-Flower Associations in High Arctic with Special Reference to Nectar." *Oikos* 19: 359–387.

Hoefer, C. 2011. "Time and Chance Propensities." In *The Oxford Handbook of Philosophy of Time*, ed. C. Callender. Oxford, UK: Oxford University Press, 68–90.

Howerton, C. L., and J. A. Mench. 2014. "Running around the Clock: Competition, Aggression and Temporal Partitioning of Running Wheel Use in Male Mice." *Animal Behaviour* 90: 221–227.

Høye, T. T., A. Eskildsen, R. R. Hansen, J. J. Bowden, N. M. Schmidt, and W. D. Kissling. 2014. "Phenology of High-arctic Butterflies and Their Floral Resources: Species-specific Responses to Climate Change." *Current Zoology* 60: 243–251.

Høye, T. T., E. Post, H. Meltofte, N. M. Schmidt, and M. C. Forchhammer. 2007. "Rapid Advancement of Spring in the High Arctic." *Current Biology* 17: R449-R451.

Høye, T. T., E. Post, N. M. Schmidt, K. Trojelsgaard, and M. C. Forchhammer. 2013. "Shorter Flowering Seasons and Declining Abundance of Flower Visitors in a Warmer Arctic." *Nature Climate Change* 3: 759–763.

Hudson, I. L., S. W. Kim, and M. R. Keatley. 2010. "Climatic Influences on the Flowering Phenology of Four Eucalypts: A GAMLSS Approach." 18th World IMACS/MODSIM Congress, Cairns, Australia, July 13–17, 2009, 2611—2617.

Husby, A., L. E. B. Kruuk, and M. E. Visser. 2009. "Decline in the Frequency and Benefits of Multiple Brooding in Great Tits as a Consequence of a Changing Environment." *Proceedings of the Royal Society B—Biological Sciences* 276: 1845–1854.

Hutchinson, G. E. 1957. "Concluding Remarks. Population Studies, Animal Ecology, and Demography." *Cold Spring Harbor Symposia on Quantitative Biology* 22: 415–427.

Iler, A. M., T. T. Hoye, D. W. Inouye, and N. M. Schmidt. 2013a. "Long-term Trends Mask Variation in the Direction and Magnitude of Short-term Phenological Shifts." *American Journal of Botany* 100: 1398–1406.

———. 2013b. "Nonlinear Flowering Responses to Climate: Are Species Approaching Their Limits of Phenological Change?" *Philosophical Transactions of the Royal Society B—Biological Sciences* 368.

Iler, A. M., D. W. Inouye, T. T. Hoye, A. J. Miller-Rushing, L. A. Burkle, and E. B. Johnston. 2013c. "Maintenance of Temporal Synchrony between Syrphid Flies and Floral Resources Despite Differential Phenological Responses to Climate." *Global Change Biology* 19: 2348–2359.

Inouye, D. W. 2008. "Effects of Climate Change on Phenology, Frost Damage, and Floral Abundance of Montane Wildflowers." *Ecology* 89: 353–362.

Inouye, D. W., B. Barr, K. B. Armitage, and B. D. Inouye. 2000. "Climate Change Is Affecting Altitudinal Migrants and Hibernating Species." *Proceedings of the National Academy of Sciences of the United States of America* 97: 1630–1633.

IPCC. 2014. *Climate Change 2014: Impacts, Adaptation, and Vulnerability. Part A: Global and Sectoral Aspects. Contribution of Working Group II to the Fifth Assessment Report of the Intergovernmental Panel on Climate Change.* Edited by C. B. Field, V. R. Barros, D. J. Dokken, K. J. Mach, M. D. Mastrandrea, T. E. Bilir, M. Chatterjee, K. L. Ebi, Y. O. Estrada, R. C. Genova, B. Girma, E. S. Kissel, A. N. Levy, S. MacCracken, P. R. Mastrandrea, and L. L. White. Cambridge, UK: Cambridge University Press.

Ishioka, R., O. Muller, T. Hiura, and G. Kudo. 2013. "Responses of Leafing Phenology and Photosynthesis to Soil Warming in Forest-floor Plants." *Acta Oecologica— International Journal of Ecology* 51: 34–41.

Iwasa, Y., and S. A. Levin. 1995. "The Timing of Life History Events." *Journal of Theoretical Biology* 172: 33–42.

Jaksic, F. M. 1982. "Inadequacy of Activity Time as a Nice Difference—The Case of Dirunal and Nocturnal Raptors." *Oecologia* 52: 171–175.

Janzen, D. H. 1967. "Synchronization of Sexual Reproduction of Trees within Dry Season in Central America." *Evolution* 21: 620–637.

Jarvinen, A. 1989. "Patterns and Causes of Long-term Variation in Reproductive Traits of the Pied Flycatcher *Ficedula hypoleuca* in Finnish Lapland." *Ornis Fennica* 66: 24–31.

Jeong, S. J., C. H. Ho, H. J. Gim, and M. E. Brown. 2011. "Phenology Shifts at Start vs. End of Growing Season in Temperate Vegetation over the Northern Hemisphere for the Period 1982–2008." *Global Change Biology* 17: 2385–2399.

Jeong, S.-J., C.-H. Ho, and J.-H. Jeong. 2009. "Increase in Vegetation Greenness and Decrease in Springtime Warming over East Asia." *Geophysical Research Letters* 36: L02710.

Johansson, J., N. P. Kristensen, J. A. Nilsson, and N. Jonzen. 2015a. The Eco-evolutionary Consequences of Interspecific Phenological Asynchrony—A Theoretical Perspective." *Oikos* 124: 102–112.

Johansson, J., J. A. Nilsson, and N. Jonzen. 2015b. "Phenological Change and Ecological Interactions: An Introduction." *Oikos* 124: 1–3.

John, C. L. 2016. *Against the Spring Wave: Ungulate Migration Phenology in a Changing Arctic.* M.S. Thesis, Pennsylvania State University.

Jonsson, A. M., G. Appelberg, S. Harding, and L. Barring. 2009. "Spatio-temporal Impact of Climate Change on the Activity and Voltinism of the Spruce Bark Beetle, *Ips Typographus.*" *Global Change Biology* 15: 486–499.

Julien, Y., and J. A. Sobrino. 2009. "Global Land Surface Phenology Trends from GIMMS Database." *International Journal of Remote Sensing* 30: 3495–3513.

Kalapanida, M., and P. V. Petrakis. 2012. "Temporal Partitioning in an Assemblage of Insect Defoliators Feeding on Oak on a Mediterranean Mountain." *European Journal of Entomology* 109: 55–69.

Kant, I. 1781. *Critique of Pure Reason.* Cambridge, UK: Cambridge University Press.

Karlsson, B. 2014. "Extended Season for Northern Butterflies." *Journal of Biometeorology* 58: 691–701.

Kashkarov, D., and V. Kurbatov. 1930. "Preliminary Ecological Survey of the Vertebrate Fauna of the Central Kara-Kum Desert in West Turkestan." *Ecology* 11: 35–60.

Kearns, C. A., D. W. Inouye, and N. M. Waser. 1998. "Endangered Mutualisms: The Conservation of Plant-Pollinator Interactions." *Annual Review of Ecology and Systematics* 29: 83–112.

Keatley, M. R., T. D. Fletcher, I. L. Hudson, and P. K. Ades. 2002. "Phenological Studies in Australia: Potential Application in Historical and Future Climate Analysis." *International Journal of Climatology* 22: 1769–1780.

Kelly, C. K., M. G. Bowler, G. A. Fox, C. K. Kelly, M. G. Bowler, and G. A. Fox, eds. 2013. *Temporal Dynamics and Ecological Processes*. Cambridge, UK: Cambridge University Press.

Kerby, J., and E. Post. 2013a. "Capital and Income Breeding Traits Differentiate Trophic Match-Mismatch Dynamics in Large Herbivores." *Philosophical Transactions of the Royal Society B—Biological Sciences* 368: 20120484.

———. 2013b. "Advancing Plant Phenology and Reduced Herbivore Production in a Terrestrial System Associated with Sea Ice Decline." *Nature Communications* 4: 2514.

Klein, A.-M., B. E. Vaissiere, J. H. Cane, I. Steffan-Dewenter, S. A. Cunningham, C. Kremen, and T. Tscharntke. 2007. "Importance of Pollinators in Changing Landscapes for World Crops." *Proceedings of the Royal Society B—Biological Sciences* 274: 303–313.

Koh, I., E. V. Lonsdorf, N. M. Williams, C. Brittain, R. Isaacs, J. Gibbs, and T. H. Ricketts. 2016. "Modeling the Status, Trends, and Impacts of Wild Bee Abundance in the United States." *Proceedings of the National Academy of Sciences of the United States of America* 113: 140–145.

Kraft, N.J.B., and D. D. Ackerly. 2010. "Functional Trait and Phylogenetic Tests of Community Assembly across Spatial Scales in an Amazonian Forest." *Ecological Monographs* 80: 401–422.

Kremen, C., N. M. Williams, M. A. Aizen, B. Gemmill-Herren, G. LeBuhn, R. Minckley, L. Packer, S. G. Potts, T. A. Roulston, I. Steffan-Dewenter, D. P. Vazquez, R. Winfree, L. Adams, E. E. Crone, S. S. Greenleaf, T. H. Keitt, A.-M. Klein, J. Regetz, and T. H. Ricketts. 2007. "Pollination and Other Ecosystem Services Produced by Mobile Organisms: A Conceptual Framework for the Effects of Land-use Change." *Ecology Letters* 10: 299–314.

Kronfeld-Schor, N., and T. Dayan. 2003. "Partitioning of Time as an Ecological Resource." *Annual Review of Ecology Evolution and Systematics* 34: 153–181.

Kudo, G. 2014. "Vulnerability of Phenological Synchrony between Plants and Pollinators in an Alpine Ecosystem." *Ecological Research* 29: 571–581.

Kudo, G., and T. Y. Ida. 2013. "Early Onset of Spring Increases the Phenological Mismatch between Plants and Pollinators." *Ecology* 94: 2311–2320.

Kuhn, T. S. 1962. *The Structure of Scientific Revolutions*. Chicago: University of Chicago Press.

Kusano, T., and M. Inoue. 2008. "Long-term Trends toward Earlier Breeding of Japanese Amphibians." *Journal of Herpetology* 42: 608–614.

Kutz, S. J., S. Checkley, G. G. Verocai, M. Dumond, E. P. Hoberg, R. Peacock, J. P. Wu, K. Orsel, K. Seegers, A. L. Warren, and A. Abrams. 2013. "Invasion, Establishment, and Range Expansion of Two Parasitic Nematodes in the Canadian Arctic." *Global Change Biology* 19: 3254–3262.

Kutz, S. J., E. P. Hoberg, and L. Polley. 2001. "*Umingmakstrongylus pallikuukensis* (Nematoda: *Protostrongylidae*) in Gastropods: Larval Morphology, Morphometrics, and Development Rates." *Journal of Parasitology* 87: 527–535.

Lack, D. 1966. *Population Studies of Birds*. Oxford, UK: Clarendon Press.

Lambert, A. M., A. J. Miller-Rushing, and D. W. Inouye. 2010. "Changes in Snowmelt Date and Summer Precipitation Affect the Flowering Phenology of *Erythronium grandiflorum* (Glacier Lily: *Liliaceae*)." *American Journal of Botany* 97: 1431–1437.

Lane, J. E., L.E.B. Kruuk, A. Charmantier, J. O. Murie, and F. S. Dobson. 2012. "Delayed Phenology and Reduced Fitness Associated with Climate Change in a Wild Hibernator." *Nature* 489: 554–558.

Laube, J., T. H. Sparks, N. Estrella, J. Hofler, D. P. Ankerst, and A. Menzel. 2014. "Chilling Outweighs Photoperiod in Preventing Precocious Spring Development." *Global Change Biology* 20: 170–182.

Law, B., C. Mackowski, L. Schoer, and T. Tweedie. 2000. "Flowering Phenology of Myrtaceous Trees and Their Relation to Climatic, Environmental and Disturbance Variables in Northern New South Wales." *Austral Ecology* 25: 160–178.

Lee, S. D., E. R. Ellwood, S. Y. Park, and R. B. Primack. 2011. "Late-arriving Barn Swallows Linked to Population Declines." *Biological Conservation* 144: 2182–2187.

Levine, J. M., and C. M. D'Antonio. 1999. "Elton Revisited: A Review of Evidence Linking Diversity and Invasibility." *Oikos* 87: 15–26.

Liancourt, P., L. A. Spence, B. Boldgiv, A. Lkhagva, B. R. Helliker, B. B. Casper, and P. S. Petraitis. 2012. "Vulnerability of the Northern Mongolian Steppe to Climate Change: Insights from Flower Production and Phenology." *Ecology* 93: 815–824.

Lieth, H., ed. 1974. *Phenology and Seasonality Modeling*. New York: Springer.

Lima, S. L. 1998. "Nonlethal Effects in the Ecology of Predator-Prey Interactions—What Are the Ecological Effects of Anti-predator Decision-making?" *Bioscience* 48: 25–34.

Lima, S. L., and L. M. Dill. 1990. "Behavioral Decisions Made under the Risk of Predation—A Review and Prospectus." *Canadian Journal of Zoology—Revue Canadienne de Zoologie* 68: 619–640.

Liu, Y. Z., P. B. Reich, G. Y. Li, and S. C. Sun. 2011. "Shifting Phenology and Abundance under Experimental Warming Alters Trophic Relationships and Plant Reproductive Capacity." *Ecology* 92: 1201–1207.

Loescher, H. W., S. F. Oberbauer, H. L. Gholz, and D. B. Clark. 2003. "Environmental Controls on Net Ecosystem-level Carbon Exchange and Productivity in a Central American Tropical Wet Forest." *Global Change Biology* 9: 396–412.

Lu, W. Q., Q. J. Meng, N.J.C. Tyler, K. A. Stokkan, and A.S.I. Loudon. 2010. "A Circadian Clock Is Not Required in an Arctic Mammal." *Current Biology* 20: 533–537.

Ludwichowski, I. 1997. "Long-term Changes of Wing-length, Body Mass and Breeding Parameters in First-time Breeding Females of Goldeneyes (*Bucephala clangula clangula*) in Northern Germany." *Vogelwarte* 39: 103–116.

Ma, T., and C. G. Zhou. 2012. "Climate-associated Changes in Spring Plant Phenology in China." *International Journal of Biometeorology* 56: 269–275.

MacArthur, R. H. 1957. "On the Relative Abundance of Bird Species." *Proceedings of the National Academy of Sciences of the United States of America* 43: 293–295.

MacArthur, R. H., and R. Levins. 1964. "Competition, Habitat Selection, and Character Displacement in a Patchy Environment." *Proceedings of the National Academy of Sciences of the United States of America* 51: 1207–1210.

MacArthur, R. H., and R. Levins. 1967. "The Limiting Similarity, Convergence, and Divergence of Coexisting Species." *American Naturalist* 101: 377–385.

MacArthur, R. H., and E. O. Wilson. 1967. *The Theory of Island Biogeography.* Princeton, NJ: Princeton University Press.

Macinnes, C. D., E. H. Dunn, D. H. Rusch, F. Cooke, and F. G. Cooch. 1990. "Advancement of Goose Nesting Dates in the Hudson Bay Region 1951–1986." *Canadian Field-Naturalist* 104: 295–297.

Mackas, D. L., R. Goldblatt, and A. G. Lewis. 1998. "Interdecadal Variation in Developmental Timing of Neocalanus Plumchrus Populations at Ocean Station P in the Subarctic North Pacific." *Canadian Journal of Fisheries and Aquatic Sciences* 55: 1878–1893.

Mahall, B. E., and F. H. Bormann. 1978. "Quantitative Description of the Vegetative Phenology of Herbs in a Northern Hardwood Forest." *Botanical Gazette* 139: 467–481.

Marchin, R. M., C. F. Salk, W. A. Hoffmann, and R. R. Dunn. 2015. "Temperature Alone Does Not Explain Phenological Variation of Diverse Temperate Plants under Experimental Warming." *Global Change Biology* 21: 3138–3151.

Marino, G. P., D. P. Kaiser, L. Gu, and D. M. Ricciuto. 2011. "Reconstruction of False Spring Occurrences over the Southeastern United States, 1901–2007: An Increasing Risk of Spring Freeze Damage?" *Environmental Research Letters* 6: 024015.

May, R. M. 1974. "Time-delay versus Stability in Population Models with Two and Three Trophic Levels." *Ecology* 54: 315–325.

———. 1975. "Deterministic Models with Chaotic Dynamics." *Nature* 256: 165–166.

May, R. M., and R. M. Anderson. 1979. "Population Biology of Infectious Diseases." *Nature* 280: 455–461.

May, R. M., and R. H. MacArthur. 1972. "Niche Overlap as a Function of Environmental Variability." *Proceedings of the National Academy of Sciences of the United States of America* 69: 1109–1113.

McCleery, R. H., and C. M. Perrins. 1998. ". . . Temperature and Egg-laying Trends. *Nature* 391: 30–31.

McKinney, A. M., P. J. CaraDonna, D. W. Inouye, B. Barr, C. D. Bertelsen, and N. M. Waser. 2012. "Asynchronous Changes in Phenology of Migrating Broad-tailed Hummingbirds and Their Early-season Nectar Resources." *Ecology* 93: 1987–1993.

McKinney, A. M., and K. Goodell. 2011. "Plant-Pollinator Interactions between an Invasive and Native Plant Vary between Sites with Different Flowering Phenology." *Plant Ecology* 212: 1025–1035.

McTaggart, J.M.E. 1908. "The Unreality of Time." *Mind* 17: 457–474.

Meldgaard, M. 1986. "The Greenland Caribou— Zoogeography, Taxonomy, and Population Dynamics." *Meddelelser Om Grønland* 20: 1–88.

Memmott, J., P. G. Craze, N. M. Waser, and M. V. Price. 2007. "Global Warming and the Disruption of Plant-Pollinator Interactions." *Ecology Letters* 10: 710–717.

Menzel, A., N. Estrella, and P. Fabian. 2001. "Spatial and Temporal Variability of the Phenological Seasons in Germany from 1951 to 1996." *Global Change Biology* 7: 657–666.

Menzel, A., and P. Fabian. 1999. "Growing Season Extended in Europe." *Nature* 397: 659–659.

Meyer, U. 2013. *The Nature of Time.* Oxford, UK: Oxford University Press.

Mickelson, M. J., P. Dann, and J. M. Cullen. 1992. "Sea Temperature in Bass Strait and Breeding Success of the Little Penguin Eudyptula Minor at Phillip Island, Southeastern Australia." *Emu* 91: 355–368.

Mills, L. S., M. Zimova, J. Oyler, S. Running, J. T. Abatzoglou, and P. M. Lukacs. 2013. "Camouflage Mismatch in Seasonal Coat Color Due to Decreased Snow Duration." *Proceedings of the National Academy of Sciences of the United States of America* 110: 7360–7365.

Minkowski, H. 1909. "Space and Time." *Physikalische Zeitschrift* 10: 75–88.

Mitton, J. B., and S. M. Ferrenberg. 2012. "Mountain Pine Beetle Develops an Unprecedented Summer Generation in Response to Climate Warming." *American Naturalist* 179: E163–E171.

Molau, U. 1997. "Responses to Natural Climatic Variation and Experimental Warming in Two Tundra Plant Species with Contrasting Life Forms: *Cassiope tetragona* and *Ranunculus nivalis*." *Global Change Biology* 3: 97–107.

Molau, U., U. Nordenhall, and B. Eriksen. 2005. "Onset of Flowering and Climate Variability in an Alpine Landscape: A 10-year Study from Swedish Lapland." *American Journal of Botany* 92: 422–431.

Monroe, A. P., K. K. Hallinger, R. L. Brasso, and D. A. Cristol. 2008. "Occurrence and Implications of Double Brooding in a Southern Population of Tree Swallows." *Condor* 110: 382–386.

Moore, P. J., R. C. Thompson, and S. J. Hawkins. 2011. "Phenological Changes in Intertidal Con-specific Gastropods in Response to Climate Warming." *Global Change Biology* 17: 709–719.

Mulder, C. P. H., D. T. Iles, and R. F. Rockwell. 2017. "Increased Variance in Temperature and Lag Effects Alter Phenological Responses to Rapid Warming in a Subarctic Plant Community." *Global Change Biology* 23: 801–814.

Munguia-Rosas, M. A., J. Ollerton, V. Parra-Tabla, and J. Arturo De-Nova. 2011. "Meta-analysis of Phenotypic Selection on Flowering Phenology Suggests That Early Flowering Plants Are Favoured." *Ecology Letters* 14: 511–521.

Musolin, D. L. 2007. "Insects in a Warmer World: Ecological, Physiological and Life-history Responses of True Bugs (Heteroptera) to Climate Change." *Global Change Biology* 13: 1565–1585.

Myneni, R. B., C. D. Keeling, C. J. Tucker, G. Asrar, and R. R. Nemani. 1997. "Increased Plant Growth in the Northern High Latitudes from 1981 to 1991." *Nature* 386: 698–702.

Neeman, N., N. J. Robinson, F. V. Paladino, J. R. Spotila, and M. P. O'Connor. 2015. "Phenology Shifts in Leatherback Turtles (*Dermochelys Coriacea*) Due to Changes in Sea Surface Temperature." *Journal of Experimental Marine Biology and Ecology* 462: 113–120.

Newton, I. 1687. *Philosophiae Naturalis Principia Mathematica*. London.

NOAA. 2016. *Global Analysis—November 2015: 2015 Year-to-Date Temperatures Versus Previous Years*. Washington, DC: National Oceanic and Atmospheric Association—National Centers for Environmental Information.

Numata, S., M. Yasuda, T. Okuda, N. Kachi, and N.S.M. Noor. 2003. "Temporal and Spatial Patterns of Mass Flowerings on the Malay Peninsula." *American Journal of Botany* 90: 1025–1031.

Opler, P. A., G. W. Frankie, and H. G. Baker. 1980. "Comparative Phenological Studies of Treelet and Shrub Species in Tropical Wet and Dry Forests in the Lowlands of Costa Rica." *Journal of Ecology* 68: 167–188.

Orme, C. D. L., R. G. Davies, M. Burgess, F. Eigenbrod, N. Pickup, V. A. Olson, A. J. Webster, T. S. Ding, P. C. Rasmussen, R. S. Ridgely, A. J. Stattersfield, P. M.

Bennett, T. M. Blackburn, K. J. Gaston, and I.P.F. Owens. 2005. "Global Hotspots of Species Richness Are Not Congruent with Endemism or Threat." *Nature* 436: 1016–1019.

Osborne, C. P., I. Chuine, D. Viner, and F. I. Woodward. 2000. "Olive Phenology as a Sensitive Indicator of Future Climatic Warming in the Mediterranean." *Plant Cell and Environment* 23: 701–710.

Ovaskainen, O., S. Skorokhodova, M. Yakovleva, A. Sukhov, A. Kutenkov, N. Kutenkova, A. Shcherbakov, E. Meyke, and M. D. Delgado. 2013. "Community-level Phenological Response to Climate Change." *Proceedings of the National Academy of Sciences of the United States of America* 110: 13434–13439.

Ozgul, A., D. Z. Childs, M. K. Oli, K. B. Armitage, D. T. Blumstein, L. E. Olson, S. Tuljapurkar, and T. Coulson. 2010. "Coupled Dynamics of Body Mass and Population Growth in Response to Environmental Change." *Nature* 466: 482–U485.

Pace, M. L., J. J. Cole, S. R. Carpenter, and J. F. Kitchell. 1999. "Trophic Cascades Revealed in Diverse Ecosystems." *Trends in Ecology and Evolution* 14: 483–488.

Pachauri, R. K., and A. Reisinger, eds. 2007. *Climate Change 2007: Synthesis Report. Contribution of Working Groups I, II, and III to the Fourth Assessment Report of the Intergovernmental Panel on Climate Change.* Geneva: IPCC.

Park, J. S. 2017. "A Race against Time: Habitat Alteration by Snow Geese Prunes the Seasonal Sequence of Mosquito Emergence in a Subarctic Brackish Landscape." *Polar Biology* 40: 553–561.

Parmesan, C. 2006. "Ecological and Evolutionary Responses to Recent Climate Change." *Annual Review of Ecology Evolution and Systematics* 37: 637–669.

———. 2007. "Influences of Species, Latitudes and Methodologies on Estimates of Phenological Response to Global Warming." *Global Change Biology* 13: 1860–1872.

Parmesan, C., and G. Yohe. 2003. "A Globally Coherent Fingerprint of Climate Change Impacts across Natural Systems." *Nature* 421: 37–42.

Pau, S., E. M. Wolkovich, B. I. Cook, T. J. Davies, N.J.B. Kraft, K. Bolmgren, J. L. Betancourt, and E. E. Cleland. 2011. "Predicting Phenology by Integrating Ecology, Evolution and Climate Science." *Global Change Biology* 17: 3633–3643.

Pau, S., E. M. Wolkovich, B. I. Cook, C. J. Nytch, J. Regetz, J. K. Zimmerman, and S. J. Wright. 2013. "Clouds and Temperature Drive Dynamic Changes in Tropical Flower Production." *Nature Climate Change* 3: 838–842.

Pedersen, C., and E. Post. 2008. "Interactions between Herbivory and Warming in Aboveground Biomass Production of Arctic Vegetation." *BMC Ecology* 8: 17.

Penrose, R. 1996a. "Quantum Theory and Spacetime." In *The Nature of Space and Time*, ed. S. Hawking and R. Penrose. Princeton, NJ: Princeton University Press, 61–74.

———. 1996b. "Structure of Spacetime Singularities." In *The Nature of Space and Time*, ed. S. Hawking and R. Penrose. Princeton, NJ: Princeton University Press, 27–36.

Penuelas, J., I. Filella, and P. Comas. 2002. "Changed Plant and Animal Life Cycles from 1952 to 2000 in the Mediterranean Region." *Global Change Biology* 8: 531–544.

Peres, C. A. 1994. "Primate Responses to Phenological Changes in an Amazonian Terra-firme Forest." *Biotropica* 26: 98–112.

Petanidou, T., A. S. Kallimanis, S. P. Sgardelis, A. D. Mazaris, J. D. Pantis, and N. M. Waser. 2014. "Variable Flowering Phenology and Pollinator Use in a Community Suggest Future Phenological Mismatch." *Acta Oecologica—International Journal of Ecology* 59: 104–111.

Peterson, A. G., and J. T. Abatzoglou. 2014. "Observed Changes in False Springs over the Contiguous United States." *Geophysical Research Letters* 41: 2156–2162.

Phillimore, A. B., S. Stalhandske, R. J. Smithers, and R. Bernard. 2012. "Dissecting the Contributions of Plasticity and Local Adaptation to the Phenology of a Butterfly and Its Host Plants." *American Naturalist* 180: 655–670.

Pianka, E. R. 1970. "R-Selection and K-Selection." *American Naturalist* 104: 592–597.

Piao, S. L., J. Y. Fang, L. M. Zhou, P. Ciais, and B. Zhu. 2006. "Variations in Satellite-derived Phenology in China's Temperate Vegetation." *Global Change Biology* 12: 672–685.

Pimm, S. L., and J. H. Brown. 2004. "Domains of Diversity." *Science* 304: 831–833.

Pittendrigh, C. S. 1960. "Circadian Rhythms and the Circadian Organization of Living Systems." *Cold Spring Harbor Symposia on Quantitative Biology* 25: 159–184.

Plard, F., J. M. Gaillard, T. Coulson, A.J.M. Hewison, D. Delorme, C. Warnant, and C. Bonenfant. 2014. "Mismatch between Birth Date and Vegetation Phenology Slows the Demography of Roe Deer." *PLoS Biology* 12(4): e1001828.

Pleasants, J. M. 1980. "Competition for Bumblebee Pollinators in Rocky Mountain Plant Communities." *Ecology* 61: 1446–1459.

Pollard, E. 1991. "Changes in the Flight of the Hedge Brown Butterfly *Pyronia tithonus* during Range Expansion." *Journal of Animal Ecology* 60: 737–748.

Poole, R. W., and B. J. Rathcke. 1979. "Regularity, Randomness, and Aggregation in Flowering Phenologies." *Science* 203: 470–471.

Post, E. 2013. "Ecology of Climate Change: The Importance of Biotic Interactions." Monographs in Population Biology No. 52. Princeton, NJ: Princeton University Press.

Post, E., and M. Avery. In Press. "Phenological Dynamics in Pollinator-Plant Associations Related to Climate Change." In *Climate Change and Biodiversity*, ed. T. E. Lovejoy and L. Hannah. New Haven, CT: Yale University Press.

Post, E., P. S. Bøving, C. Pedersen, and M. A. MacArthur. 2003. "Synchrony between Caribou Calving and Plant Phenology in Depredated and Non-depredated Populations." *Canadian Journal of Zoology* 81: 1709–1714.

Post, E., and J. Brodie. 2015. "Anticipating Novel Conservation Risks of Increased Human Access to Remote Regions with Warming." *Climate Change Responses* 2: 2.

Post, E., and M. C. Forchhammer. 2008. "Climate Change Reduces Reproductive Success of an Arctic Herbivore through Trophic Mismatch." *Philosophical Transactions of the Royal Society of London, Series B* 363: 2369–2375.

Post, E., M. C. Forchhammer, M. S. Bret-Harte, T. V. Callaghan, T. R. Christensen, B. Elberling, A. D. Fox, O. Gilg, D. S. Hik, T. T. Hoye, R. A. Ims, E. Jeppesen, D. R. Klein, J. Madsen, A. D. McGuire, S. Rysgaard, D. E. Schindler, I. Stirling, M. P. Tamstorf, N.J.C. Tyler, R. Van Der Wal, J. M. Welker, P. A. Wookey, N. M. Schmidt, and P. Aastrup. 2009. "Ecological Dynamics across the Arctic Associated with Recent Climate Change." *Science* 325: 1355–1358.

Post, E., M. C. Forchhammer, and N. C. Stenseth. 1999a. "Population Ecology and the North Atlantic Oscillation (NAO)." *Ecological Bulletins* 47: 117–125.

Post, E., M. C. Forchhammer, N. C. Stenseth, and T. V. Callaghan. 2001a. "The Timing of Life-history Events in a Changing Climate." *Proceedings of the Royal Society of London, Series B* 268: 15–23.

Post, E., J. Kerby, C. Pedersen, and H. Steltzer. 2016. "Highly Individualistic Rates of Plant Phenological Advance Associated with Arctic Sea Ice Dynamics." *Biology Letters* 12: 20160332.

Post, E., and D. R. Klein. 1999. "Caribou Calf Production and Seasonal Range Quality during a Population Decline." *Journal of Wildlife Management* 63: 335–345.

Post, E., S. A. Levin, Y. Iwasa, and N. C. Stenseth. 2001b. "Reproductive Asynchrony Increases with Environmental Disturbance." *Evolution* 55: 830–834.

Post, E., and C. Pedersen. 2008. "Opposing Plant Community Responses to Warming with and without Herbivores." *Proceedings of the National Academy of Sciences* 105: 12353–12358.

Post, E., C. Pedersen, C. C. Wilmers, and M. C. Forchhammer. 2008a. "Phenological Sequences Reveal Aggregate Life History Response to Climatic Warming." *Ecology* 89: 363–370.

———. 2008b. "Warming, Plant Phenology, and the Spatial Dimension of Trophic Mismatch for Large Herbivores." *Proceedings of the Royal Society of London, Series B* 275: 2005–2013.

Post, E., R. O. Peterson, N. C. Stenseth, and B. E. McLaren. 1999b. "Ecosystem Consequences of Wolf Behavioural Response to Climate." *Nature* 401: 905–907.

Post, E., B. A. Steinman, and M. E. Mann. 2018. "Acceleration of Phenological Advance and Warming with Latitude over the Past Century." *Scientific Reports* 8: 3297.

Post, E., and N. C. Stenseth. 1998. "Large-scale Climatic Fluctuation and Population Dynamics of Moose and White-tailed Deer." *Journal of Animal Ecology.*

———. 1999. "Climatic Variability, Plant Phenology, and Northern Ungulates." *Ecology* 80: 1322–1339.

Post, E. S. 1995. "Comparative Foraging Ecology and Social Dynamics of Caribou (*Rangifer Tarandus*)." Ph.D. Dissertation, University of Alaska, Fairbanks.

Potter, S., M. Cabbage, and L. McCarthy. 2017. *NASA, NOAA Data Show 2016 Warmest Year on Record Globally.* Washington, DC: NASA.

Potts, S. G., J. C. Biesmeijer, C. Kremen, P. Neumann, O. Schweiger, and W. E. Kunin. 2010a. "Global Pollinator Declines: Trends, Impacts and Drivers." *Trends in Ecology and Evolution* 25: 345–353.

Potts, S. G., S.P.M. Roberts, R. Dean, G. Marris, M. S. Brown, R. J. Jones, P. Neumann, and J. Settele. 2010b. "Declines of Managed Honey Bees and Beekeepers in Europe." *Journal of Apicultural Research* 49: 15–22.

Power, M. E. 1992. "Top-down and Bottom-up Forces in Food Webs—Do Plants Have Primacy?" *Ecology* 73: 733–746.

Power, M. E., W. J. Matthews, and A. J. Stewart. 1985. "Grazing Minnows, Piscivorous Bass, and Stream Algae—Dynamics of a Strong Interaction." *Ecology* 66: 1448–1456.

Price, H. 2011. "The Flow of Time." In *The Oxford Handbook of Philosophy of Time*, ed. C. Callender. Oxford, UK: Oxford University Press, 276–311.

Prieto, P., J. Penuelas, U. Niinemets, R. Ogaya, I. K. Schmidt, C. Beier, A. Tietema, A. Sowerby, B. A. Emmett, E. K. Lang, G. Kroel-Dulay, B. Lhotsky, C. Cesaraccio, G. Pellizzaro, G. De Dato, C. Sirca, and M. Estiarte. 2009. "Changes in the Onset of Spring Growth in Shrubland Species in Response to Experimental Warming along a North-South Gradient in Europe." *Global Ecology and Biogeography* 18: 473–484.

Primack, R. B., I. Ibanez, H. Higuchi, S. D. Lee, A. J. Miller-Rushing, A. M. Wilson, and J. A. Silander. 2009. "Spatial and Interspecific Variability in Phenological Responses to Warming Temperatures." *Biological Conservation* 142: 2569–2577.

Radville, L., M. L. McCormack, E. Post, and D. M. Eissenstat. 2016. "Root Phenology in a Changing Climate." *Journal of Experimental Botany* 67 (12): 3617–3628.

Rafferty, N. E., P. J. CaraDonna, and J. L. Bronstein. 2015. "Phenological Shifts and the Fate of Mutualisms." *Oikos* 124: 14–21.

Rafferty, N. E., P. J. CaraDonna, L. A. Burkle, A. M. Iler, and J. L. Bronstein. 2013. "Phenological Overlap of Interacting Species in a Changing Climate: An Assessment of Available Approaches." *Ecology and Evolution* 3: 3183–3193.

Rafferty, N. E., and A. R. Ives. 2011. "Effects of Experimental Shifts in Flowering Phenology on Plant-Pollinator Interactions." *Ecology Letters* 14: 69–74.

———. 2012. "Pollinator Effectiveness Varies with Experimental Shifts in Flowering Time." *Ecology* 93: 803–814.

Ramos, D. M., P. Diniz, and J.F.M. Valls. 2014. "Habitat Filtering and Interspecific Competition Influence Phenological Diversity in an Assemblage of Neotropical Savanna Grasses." *Brazilian Journal of Botany* 37: 29–36.

Raney, E. C., and E. A. Lachner. 1946. "Age, Growth, and Habits of the Hog Sucker, *Hypentelium nigricans* (*LeSueur*), in New York." *American Midland Naturalist* 36: 76–86.

Ranta, E., V. Kaitala, and P. Lundberg. 1997. "The Spatial Dimension in Population Fluctuations." *Science* 278: 1621–1623.

Ranta, E., V. Kaitala, and P. Lundberg. 1998. "Population Variability in Space and Time: The Dynamics of Synchronous Population Fluctuations." *Oikos* 83: 376–382.

Rathcke, B. 1988. "Flowering Phenologies in a Shrub Community—Competition and Constraints." *Journal of Ecology* 76: 975–994.

Rawal, D. S., S. Kasel, M. R. Keatley, C. Aponte, and C. R. Nitschke. 2014. "Environmental Effects on Growth Phenology of Co-occurring Eucalyptus Species." *International Journal of Biometeorology* 58: 427–442.

Rawal, D. S., S. Kasel, M. R. Keatley, and C. R. Nitschke. 2015. "Herbarium Records Identify Sensitivity of Flowering Phenology of Eucalypts to Climate: Implications for Species Response to Climate Change." *Austral Ecology* 40: 117–125.

Reading, C. J. 1998. "The Effect of Winter Temperatures on the Timing of Breeding Activity in the Common Toad Bufo Bufo." *Oecologia* 117: 469–475.

Reale, D., A. G. McAdam, S. Boutin, and D. Berteaux. 2003. "Genetic and Plastic Responses of a Northern Mammal to Climate Change." *Proceedings of the Royal Society of London Series B—Biological Sciences* 270: 591–596.

Reich, P. B., M. B. Walters, and D. S. Ellsworth. 1997. "From Tropics to Tundra: Global Convergence in Plant Functioning." *Proceedings of the National Academy of Sciences of the United States of America* 94: 13730–13734.

Reyes-Fox, M., H. Steltzer, M. J. Trlica, G. S. McMaster, A. A. Andales, D. R. LeCain, and J. A. Morgan. 2014. "Elevated CO_2 Further Lengthens Growing Season under Warming Conditions." *Nature* 510: 259–262.

Richards, S. A. 2002. "Temporal Partitioning and Aggression among Foragers: Modeling the Effects of Stochasticity and Individual State." *Behavioral Ecology* 13: 427–438.

Richardson, A. D., T. F. Keenan, M. Migliavacca, Y. Ryu, O. Sonnentag, and M. Toomey. 2013. "Climate Change, Phenology, and Phenological Control of Vegetation Feedbacks to the Climate System." *Agricultural and Forest Meteorology* 169: 156–173.

Ricklefs, R. E. 1987. "Community Diversity—Relative Roles of Local and Regional Processes." *Science* 235: 167–171.

———. 2011. "Applying a Regional Community Concept to Forest Birds of Eastern North America." *Proceedings of the National Academy of Sciences of the United States of America* 108: 2300–2305.

Robertson, T. B. 1923. "Consciousness and the Sense of Time." *Scientific Monthly* 16: 649–657.

Rodó, X., M. Pascual, G. Fuchs, and A.S.G. Faruque. 2002. "ENSO and Cholera: A Nonstationary Link Related to Climate Change?" *Proceedings of the National Academy of Sciences of the United States of America* 99: 12901–12906.

Roetzer, T., M. Wittenzeller, H. Haeckel, and J. Nekovar. 2000. "Phenology in Central Europe—Differences and Trends of Spring Phenophases in Urban and Rural Areas." *International Journal of Biometeorology* 44: 60–66.

Roff, D. A. 1992. *The Evolution of Life Histories.* New York: Chapman and Hall.

Root, R. B. 1967. "The Niche Exploitation Pattern of the Blue-gray Gnatcatcher." *Ecological Monographs* 37: 317–350.

Root, T. L., J. T. Price, K. R. Hall, S. H. Schneider, C. Rosenzweig, and J. A. Pounds. 2003. "Fingerprints of Global Warming on Wild Animals and Plants." *Nature* 421: 57–60.

Roy, D. B., and T. H. Sparks. 2000. "Phenology of British Butterflies and Climate Change." *Global Change Biology* 6: 407–416.

Ryg, M., and E. Jacobsen. 1982. "Seasonal Changes in Growth Rate, Feed Intake, Growth Hormone, and Thyroid Hormones in Young Male Reindeer (*Rangifer tarandus tarandus*)." *Canadian Journal of Zoology—Revue Canadienne De Zoologie* 60: 15–23.

Saino, N., D. Rubolini, E. Lehikoinen, L. V. Sokolov, A. Bonisoli-Alquati, R. Ambrosini, G. Boncoraglio, and A. P. Moller. 2009. "Climate Change Effects on Migration Phenology May Mismatch Brood Parasitic Cuckoos and Their Hosts." *Biology Letters* 5: 539–541.

Schiegg, K., G. Pasinelli, J. R. Walters, and S. J. Daniels. 2002. "Inbreeding and Experience Affect Response to Climate Change by Endangered Woodpeckers." *Proceedings of the Royal Society B—Biological Sciences* 269: 1153–1159.

Schmidt-Nielsen, K. 1984. *Scaling: Why Is Animal Size So Important?* Cambridge, UK: Cambridge University Press.

Schmitz, O. J. 1998. "Direct and Indirect Effects of Predation and Predation Risk in Old-field Interaction Webs." *American Naturalist* 151: 327–342.

———. 2004. "Perturbation and Abrupt Shift in Trophic Control of Biodiversity and Productivity." *Ecology Letters* 7: 403–409.

———. 2008. "Effects of Predator Hunting Mode on Grassland Ecosystem Function." *Science* 319: 952–954.

———. 2010. *Resolving Ecosystem Complexity.* Princeton, NJ: Princeton University Press.

Schmitz, O. J., P. A. Hamback, and A. P. Beckerman. 2000. "Trophic Cascades in Terrestrial Systems: A Review of the Effects of Carnivore Removals on Plants." *American Naturalist* 155: 141–153.

Schmitz, O. J., E. Post, C. E. Burns, and K. M. Johnston. 2003. "Ecosystem Responses to Global Climate Change: Moving beyond Color Mapping." *Bioscience* 53: 1199–1205.

Schoener, T. W. 1970. "Nonsynchronous Spatial Overlap of Lizards in Patchy Habitats." *Ecology* 51: 408–418.

———. 1974. "Resource Partitioning in Ecological Communities." *Science* 185: 27–39.

Schultze, F.E.O. 1908. "Contribution to the Psychology of Time Consciousness." *Archiv fur Die Gesamte Psychologie* 13: 275–351.

Schwartz, M. D. 1998. "Green-wave Phenology." *Nature* 394: 839–840.

Schwartz, M. D., ed. 2003. *Phenology: An Integrative Environmental Science.* Dordrecht: Kluwer Academic Publishers.

Schwartz, M. D., and J. M. Caprio. 2003. "North American First Leaf and First Bloom Lilac Phenology Data." IGBP PAGES/World Data Center for Paleoclimatology, Data Contribution Series 3 2003-078. NOAA/NGDC Paleoclimatology Program, Boulder, CO.

Schwartz, M. D., and B. E. Reiter. 2000. "Changes in North American Spring." *International Journal of Climatology* 20: 929–932.

Shapiro, A. M. 1973. "Some Climatological and Phenological Aspects of the 1971 Season." *Journal of Research on the Lepidoptera* 12: 113–126.

———. 1975. "The Temporal Component of Butterfly Species Diversity." In *Ecology and Evolution of Communities*, ed. M. L. Cody and J. M. Diamond. Cambridge, MA: Harvard University Press, 181–195.

Shapiro, A. M., R. VanBuskirk, G. Kareofelas, and W. D. Patterson. 2003. "Phenofaunistics: Seasonality as a Property of Butterfly Faunas." In *Butterflies: Ecology and Evolution Taking Flight*, ed. C. L. Boggs, W. B. Watt, and P. R. Ehrlich. Chicago: University of Chicago Press, 111–147.

Shaver, G. R., and W. D. Billings. 1977. "Effects of Daylength and Temperature on Root Elongation in Tundra Graminoids." *Oecologia* 28: 57–65.

Shaver, G. R., and J. Kummerow. 1992. "Phenology, Resource Allocation, and Growth of Arctic Vascular Plants." In *Arctic Ecosystems in a Changing Climate*, ed. F. S. Chapin III, R. L. Jefferies, J. F. Reynolds, G. R. Shaver, and J. Svoboda. New York: Academic Press, 193–238.

Shea, K., and P. Chesson. 2002. "Community Ecology Theory as a Framework for Biological Invasions." *Trends in Ecology and Evolution* 17: 170–176.

Sherry, R. A., X. H. Zhou, S. L. Gu, J. A. Arnone, D. S. Schimel, P. S. Verburg, L. L. Wallace, and Y. Q. Luo. 2007. "Divergence of Reproductive Phenology under Climate Warming." *Proceedings of the National Academy of Sciences of the United States of America* 104: 198–202.

Shukla, R. P., and P. S. Ramakrishnan. 1982. "Phenology of Trees in a Sub-tropical Humid Forest in Northeastern India." *Vegetatio* 49: 103–109.

Silvertown, J., M. Franco, and J. L. Harper, eds. 1997. *Plant Life Histories: Ecology, Phylogeny and Evolution.* Cambridge, UK: Cambridge University Press.

Skelly, D. K. 1994. "Activity Level and the Susceptibility of Anuran Larvae to Predation." *Animal Behaviour* 47: 465–468.

Skelly, D. K., L. N. Joseph, H. P. Possingham, L. K. Freidenburg, T. J. Farrugia, M. T. Kinnison, and A. P. Hendry. 2007. "Evolutionary Responses to Climate Change." *Conservation Biology* 21: 1353–1355.

Slater, F. M. 1999. "First-egg Date Fluctuations for the Pied Flycatcher *Ficedula hypoleuca* in the Woodlands of Mid-Wales in the Twentieth Century." *Ibis* 141: 497–499.

Slobodchikoff, C. N., and W. C. Schulz. 1980. "Measures of Niche Overlap." *Ecology* 61: 1051–1055.

Soberon, J. 2007. "Grinnellian and Eltonian Niches and Geographic Distributions of Species." *Ecology Letters* 10: 1115–1123.

Soberon, J., and M. Nakamura. 2009. "Niches and Distributional Areas: Concepts, Methods, and Assumptions." *Proceedings of the National Academy of Sciences of the United States of America* 106: 19644–19650.

Solga, M. J., J. P. Harmon, and A. C. Ganguli. 2014. "Timing Is Everything: An Overview of Phenological Changes to Plants and Their Pollinators." *Natural Areas Journal* 34: 227–234.

Solonen, T. 2014. "Timing of Breeding in Rural and Urban Tawny Owls *Strix aluco* in Southern Finland: Effects of Vole Abundance and Winter Weather." *Journal of Ornithology* 155: 27–36.

Sørensen, T. J. 1941. "Temperature Relations and Phenology of the Northeast Greenland Flowering Plants." *Meddelelser Om Grønland* 129: 1–305.

Spano, D., C. Cesaraccio, P. Duce, and R. L. Snyder. 1999. "Phenological Stages of Natural Species and Their Use as Climate Indicators." *International Journal of Biometeorology* 42: 124–133.

Sparks, T. H. 1999. "Phenology and the Changing Pattern of Bird Migration in Britain." *International Journal of Biometeorology* 42: 134–138.

Sparks, T. H., and O. Braslavska. 2001. "The Effects of Temperature, Altitude and Latitude on the Arrival and Departure Dates of the Swallow *Hirundo rustica* in the Slovak Republic." *International Journal of Biometeorology* 45: 212–216.

Sparks, T. H., M. Gorska-Zajaczkowska, W. Wojtowicz, and P. Tryjanowski. 2011. "Phenological Changes and Reduced Seasonal Synchrony in Western Poland." *International Journal of Biometeorology* 55: 447–453.

Sparks, T. H., K. Huber, R. L. Bland, H. Q. P. Crick, P. J. Croxton, J. Flood, R. G. Loxton, C. F. Mason, J. A. Newnham, and P. Tryjanowski. 2007. "How Consistent Are Trends in Arrival (and Departure) Dates of Migrant Birds in the UK?" *Journal of Ornithology* 148: 503–511.

Stamou, G. P., M. D. Asikidis, M. D. Argyropoulou, and S. P. Sgardelis. 1993. "Ecological Time versus Standard Clock Time—The Asymmetry of Phenologies and the Life History Strategies of Some Soil Arthropods from Mediterranean Ecosystems." *Oikos* 66: 27–35.

Stanley, D. A., M.P.D. Garratt, J. B. Wickens, V. J. Wickens, S. G. Potts, and N. E. Raine. 2015a. "Neonicotinoid Pesticide Exposure Impairs Crop Pollination Services Provided by Bumblebees." *Nature* 528: 548–550.

Stanley, D. A., K. E. Smith, and N. E. Raine. 2015b. "Bumblebee Learning and Memory Is Impaired by Chronic Exposure to a Neonicotinoid Pesticide." *Scientific Reports* 5: 16508.

Stearns, S. C. 1992. *The Evolution of Life Histories.* Oxford, UK: Oxford University Press.

Steinberg, E. K., and P. Kareiva. 1997. "Challenges and Opportunities for Empirical Evaluation of "Spatial Theory."" In *Spatial Ecology: The Role of Space in Population Dynamics and Interspecific Interactions*, ed. D. Tilman and P. Kareiva. Princeton, NJ: Princeton University Press, 318–332.

Stenseth, N. C., K. S. Chan, E. Framstad, and H. Tong. 1998a. "Phase- and Density-dependent Population Dynamics in Norwegian Lemmings: Interaction between Deterministic and Stochastic Processes." *Proceedings of the Royal Society of London Series B—Biological Sciences* 265: 1957–1968.

Stenseth, N. C., W. Falck, K. S. Chan, O. N. Bjørnstad, M. O'Donoghue, H. Tong, R. Boonstra, S. Boutin, C. J. Krebs, and N. G. Yoccoz. 1998b. "From Patterns to Processes: Phase and Density Dependencies in the Canadian Lynx Cycle." *Proceedings of the National Academy of Sciences of the United States of America* 95: 15430–15435.

Stiles, F. G. 1977. "Co-adapted Competitors—Flowering Seasons of Hummingbird-pollinated Plants in a Tropical Forest." *Science* 198: 1177–1178.

Stöckli, R., and P. L. Vidale. 2004. "European Plant Phenology and Climate as Seen in a 20-year AVHRR Land-surface Parameter Dataset." *International Journal of Remote Sensing* 25: 3303–3330.

Stone, G., P. Willmer, and S. Nee. 1996. "Daily Partitioning of Pollinators in an African Acacia Community." *Proceedings of the Royal Society B—Biological Sciences* 263: 1389–1393.

Strong, D. R. 1992. "Are Trophic Cascades All Wet—Differentiation and Donor-control in Speciose Ecosystems." *Ecology* 73: 747–754.

Sugg, P., J. S. Edwards, and J. Baust. 1983. "Phenology and Life History of Belgica Antarctica, an Antarctic Midge (*Diptera, Chironomidae*)." *Ecological Entomology* 8: 105–113.

Tang, G., J. A. Arnone, P.S.J. Verburg, R. L. Jasoni, and L. Sun. 2015. "Trends and Climatic Sensitivities of Vegetation Phenology in Semiarid and Arid Ecosystems in the US Great Basin during 1982–2011." *Biogeosciences* 12: 6985–6997.

Thackeray, S. J., P. A. Henrys, D. Hemming, J. R. Bell, M. S. Botham, S. Burthe, P. Helaouet, D. G. Johns, I. D. Jones, D. I. Leech, E. B. Mackay, D. Massimino, S. Atkinson, P. J. Bacon, T. M. Brereton, L. Carvalho, T. H. Clutton-Brock, C. Duck, M. Edwards, J. M. Elliott, S.J.G. Hall, R. Harrington, J. W. Pearce-Higgins, T. T. Hoye, L.E.B. Kruuk, J. M. Pemberton, T. H. Sparks, P. M. Thompson, I. White, I. J. Winfield, and S. Wanless. 2016. "Phenological Sensitivity to Climate across Taxa and Trophic Levels." *Nature* 535: 241–245.

Thackeray, S. J., T. H. Sparks, M. Frederiksen, S. Burthe, P. J. Bacon, J. R. Bell, M. S. Botham, T. M. Brereton, P. W. Bright, L. Carvalho, T. Clutton-Brock, A. Dawson, M. Edwards, J. M. Elliott, R. Harrington, D. Johns, I. D. Jones, J. T. Jones, D. I. Leech, D. B. Roy, W. A. Scott, M. Smith, R. J. Smithers, I. J. Winfield, and S. Wanless. 2010. "Trophic Level Asynchrony in Rates of Phenological Change for Marine, Freshwater and Terrestrial Environments." *Global Change Biology* 16: 3304–3313.

Thing, H. 1984. "Feeding Ecology of the West Greenland Caribou (*Rangifer tarandus*) in the Sisimiut-Kangerlussuaq Region." *Danish Review of Game Biology* 12: 1–53.

Thomann, M., E. Imbert, C. Devaux, and P.-O. Cheptou. 2013. "Flowering Plants under Global Pollinator Decline." *Trends in Plant Science* 18: 353–359.

Thompson, D.B.A., P. S. Thompson, and D. Nethersolethompson. 1986. "Timing of Breeding and Breeding Performance in a Population of Greenshanks (Tringa Nebularia). *Journal of Animal Ecology* 55: 181–199.

Thompson, R., and R. M. Clark. 2008. "Is Spring Starting Earlier?" *Holocene* 18: 95–104.

Tilman, D. 1988. *Plant Strategies and the Dynamics and Structure of Plant Communities.* Princeton, NJ: Princeton University Press.

Tilman, D., and P. Kareiva, eds. 1997. *Spatial Ecology: The Role of Space in Population Dynamics and Interspecific Interactions.* Princeton, NJ: Princeton University Press.

Tilman, D., and C. L. Lehman. 1997. "Habitat Destruction and Species Extinction." In *Spatial Ecology: The Role of Space in Population Dynamics and Interspecific Interactions*, ed. D. Tilman and P. Kareiva. Princeton, NJ: Princeton University Press, 233–249.

Ting, S., S. Hartley, and K. C. Burns. 2008. "Global Patterns in Fruiting Seasons. *Global Ecology and Biogeography* 17: 648–657.

Tobin, P. C., S. Nagarkatti, G. Loeb, and M. C. Saunders. 2008. "Historical and Projected Interactions between Climate Change and Insect Voltinism in a Multivoltine Species." *Global Change Biology* 14: 951–957.

Todd, B. D., D. E. Scott, J.H.K. Pechmann, and J. W. Gibbons. 2011. "Climate Change Correlates with Rapid Delays and Advancements in Reproductive Timing in an Amphibian Community." *Proceedings of the Royal Society B—Biological Sciences* 278: 2191–2197.

Tryjanowski, P., M. Rybacki, and T. Sparks. 2003. "Changes in the First Spawning Dates of Common Frogs and Common Toads in Western Poland in 1978–2002." *Annales Zoologici Fennici* 40: 459–464.

Tucker, C. J., D. A. Slayback, J. E. Pinzon, S. O. Los, R. B. Myneni, and M. G. Taylor. 2001. "Higher Northern Latitude Normalized Difference Vegetation Index and Growing Season Trends from 1982 to 1999." *International Journal of Biometeorology* 45: 184–190.

Turchin, P., and A. D. Taylor. 1992. "Complex Dynamics in Ecological Time Series." *Ecology* 73: 289–305.

Tyler, N.J.C., P. Fauchald, O. Johansen, and H. R. Christiansen. 1999. "Seasonal Inappetence and Weight Loss in Female Reindeer in Winter." *Ecological Bulletins* 47: 105–116.

Tyler, N.J.C., and N. A. Øritsland. 1989. "Why Don't Svalbard Reindeer Migrate?" *Holarctic Ecology* 12: 369–376.

vanEngelsdorp, D., J. Hayes Jr., R. M. Underwood, and J. Pettis. 2008. "A Survey of Honey Bee Colony Losses in the U.S., Fall 2007 to Spring 2008." *PLoS One* 3 (12): e4071.

van Oort, B.E.H., N.J.C. Tyler, M. P. Gerkema, L. Folkow, A. S. Blix, and K.-A. Stokkan. 2005. "Circadian Organisation in Reindeer." *Nature* 438: 1095–1096.

van Oort, B.E.H., N.J.C. Tyler, M. P. Gerkema, L. Folkow, and K.-A. Stokkan. 2007. "Where Clocks Are Redundant: Weak Circadian Mechanisms in Reindeer Living under Polar Photic Conditions." *Naturwissenschaften* 94: 183–194.

van Schaik, C. P., J. W. Terborgh, and S. J. Wright. 1993. "The Phenology of Tropical Forests: Adaptive Significance and Consequences for Primary Consumers." *Annual Review of Ecology and Systematics* 24: 353–377.

Vegvari, Z., V. Bokony, Z. Barta, and G. Kovacs. 2010. "Life History Predicts Advancement of Avian Spring Migration in Response to Climate Change." *Global Change Biology* 16: 1–11.

Visser, M. E. 2008. "Keeping Up with a Warming World: Assessing the Rate of Adaptation to Climate Change." *Proceedings of the Royal Society B—Biological Sciences* 275: 649–659.

Visser, M. E., F. Adriaensen, J. H. Van Balen, J. Blondel, A. A. Dhondt, S. Van Dongen, C. Du Feu, E. V. Ivankina, A. B. Kerimov, J. De Laet, E. Matthysen, R. McCleery, M. Orell, and D. L. Thomson. 2003. "Variable Responses to Large-scale Climate Change in European Parus Populations." *Proceedings of the Royal Society B—Biological Sciences* 270: 367–372.

Visser, M. E., C. Both, and M. M. Lambrechts. 2004. "Global Climate Change Leads to Mistimed Avian Reproduction." *Advances in Ecological Research* 35: 89–110.

Visser, M. E., L.J.M. Holleman, and P. Gienapp. 2006. "Shifts in Caterpillar Biomass Phenology Due to Climate Change and Its Impact on the Breeding Biology of an Insectivorous Bird." *Oecologia* 147: 164–172.

Visser, M. E., A. J. Van Noordwijk, J. M. Tinbergen, and C. M. Lessells. 1998. "Warmer Springs Lead to Mistimed Reproduction in Great Tits (*Parus major*)." *Proceedings of the Royal Society of London Series B—Biological Sciences* 265: 1867–1870.

Walther, G. R., E. Post, P. Convey, A. Menzel, C. Parmesan, T.J.C. Beebee, J. M. Fromentin, O. Hoegh-Guldberg, and F. Bairlein. 2002. "Ecological Responses to Recent Climate Change." *Nature* 416: 389–395.

West, G. B., J. H. Brown, and B. J. Enquist. 1997. "A General Model for the Origin of Allometric Scaling Laws in Biology." *Science* 276: 122–126.

Wheelwright, N. T. 1985. "Competition for Dispersers, and the Timing of Flowering and Fruiting in a Guild of Tropical Trees." *Oikos* 44: 465–477.

Wiederholt, R., and E. Post. 2010. "Tropical Warming and the Dynamics of Endangered Primates." *Biology Letters* 6: 257–260.

Wilczek, A. M., L. T. Burghardt, A. R. Cobb, M. D. Cooper, S. M. Welch, and J. Schmitt. 2010. "Genetic and Phsyiological Bases for Phenological Responses to Current and Predicted Climates." *Philosophical Transactions of the Royal Society B—Biological Sciences* 365: 3129–3147.

Williams, G. C. 1966. *Adaptation and Natural Selection—A Critique of Some Current Evolutionary Thought*. Princeton, NJ: Princeton University Press.

Williams, J. W., and S. T. Jackson. 2007. "Novel Climates, No-analog Communities, and Ecological Surprises." *Frontiers in Ecology and the Environment* 5: 475–482.

Williams, P. H., M. B. Araujo, and P. Rasmont. 2007. "Can Vulnerability among British Bumblebee (Bombus) Species Be Explained by Niche Position and Breadth?" *Biological Conservation* 138: 493–505.

Williams, P. H., and J. L. Osborne. 2009. "Bumblebee Vulnerability and Conservation World-wide." *Apidologie* 40: 367–387.

Winkel, W., and H. Hudde. 1996. "Long-term Changes of Breeding Parameters of Nuthatches Sitta Europaea in Two Study Areas of Northern Germany." *Journal fur Ornithologie* 137: 193–202.

Wolkovich, E. M., and E. E. Cleland. 2011. "The Phenology of Plant Invasions: A Community Ecology Perspective." *Frontiers in Ecology and the Environment* 9: 287–294.

———. 2014. "Phenological Niches and the Future of Invaded Ecosystems with Climate Change." *AoB Plants* 6: plu013.

Wolkovich, E. M., B. I. Cook, and T. J. Davies. 2014a. "Progress towards an Interdisciplinary Science of Plant Phenology: Building Predictions across Space, Time and Species Diversity." *New Phytologist* 201: 1156–1162.

Wolkovich, E. M., B. I. Cook, K. K. McLauchlan, and T. J. Davies. 2014b. "Temporal Ecology in the Anthropocene." *Ecology Letters* 17: 1365–1379.

Wong, S. T., C. Servheen, L. Ambu, and A. Norhayati. 2005. "Impacts of Fruit Production Cycles on Malayan Sun Bears and Bearded Pigs in Lowland Tropical Forest of Sabah, Malaysian Borneo." *Journal of Tropical Ecology* 21: 627–639.

Wooller, R. D., K. C. Richardson, and B. G. Collins. 1993. "The Relationshipp between Nectar Supply and the Rate of Capture of a Nectar-dependent Small Marsupial *Tarsipes rostratus*." *Journal of Zoology* 229: 651–658.

Wright, S. J. 1991. "Seasonal Drought and the Phenology of Understory Shrubs in a Tropical Moist Forest." *Ecology* 72: 1643–1657.

Wright, S. J., and O. Calderon. 2006. "Seasonal, El Nino and Longer Term Changes in Flower and Seed Production in a Moist Tropical Forest." *Ecology Letters* 9: 35–44.

Wright, S. J., H. C. Muller-Landau, and J. Schipper. 2009. "The Future of Tropical Species on a Warmer Planet." *Conservation Biology* 23: 1418–1426.

Yang, L. H., and V.H.W. Rudolf. 2010. "Phenology, Ontogeny and the Effects of Climate Change on the Timing of Species Interactions." *Ecology Letters* 13: 1–10.

Zalakevicius, M., G. Bartkeviciene, L. Raudonikis, and J. Janulaitis. 2006. "Spring Arrival Response to Climate Change in Birds: A Case Study from Eastern Europe." *Journal of Ornithology* 147: 326–343.

Zimova, M., L. S. Mills, P. M. Lukacs, and M. S. Mitchell. 2014. "Snowshoe Hares Display Limited Phenotypic Plasticity to Mismatch in Seasonal Camouflage." *Proceedings of the Royal Society B—Biological Sciences* 281.

Ziv, Y., Z. Abramsky, B. P. Kotler, and A. Subach. 1993. "Interference Competition and Temporal and Habitat Partitioning in 2 Gerbil Species." *Oikos* 66: 237–246.

Index

abiotic conditions, timing of favorable, 73–74, 77, 79, 81–82, 85–93

abiotic constraints, alleviation of, 84, 112, 117, 164

absolute phenological niche, 67–71, 73, 77

absolute time, 60, 62, 67–68, 181

absolute timing, 12, 51, 153

allocation of time, 6, 9, 14, 45–46, 49–50, 54, 58, 94, 97, 99, 108, 112, 166, 169, 179

amphibians, 15, 20–23, 25–27, 33–36, 103, 133, 191–93

arrival: migratory, 15, 145; timing of, 56, 107, 143–44, 146, 149, 153–55, 157–59

arrival dates, 144, 146, 159

arrival phenology, 146, 155

arrow of time, 60–61

astrophysics, 10, 14

atmospheric moisture content, 16, 167

availability of time, 42, 50, 65, 85, 108, 130, 132, 167, 178, 181, 185

bees, 83

beetle, mountain pine, 105

biotic conditions: seasonal timing of favorable, 81, 91; timing of favorable, 73–74, 79, 81–82, 84–93

biotic interactions, 65, 183–85

biotic promoter of earlier timing, 109

bumblebees, 83, 165

butterflies, 2, 21–22, 96, 103, 105, 159, 165

Calamagrostis, 100, 120, 152

caribou, 48, 143–45, 146, 149–51, 153–55, 157–61

caribou arrival, 154, 158–59

caribou arrival phenology, 149, 154

cauchy surface, 13

changes in median dates of activity, 135, 140

climate change: implications of, 70, 106; recent, 17, 104

clock time, 50

community, vertical, 135, 140

community emergence: onset of, 128–30; timing of, 128, 130, 132

community-level emergence, 100, 120–21, 128

community-wide emergence timing, 153–54

competition, 3–4, 6, 46–49, 74–75, 104, 107, 109, 111–12, 114–15, 117, 137, 141, 143, 163–64, 177–80; density-dependent, 109, 111

competitors, conspecific and heterospecific, 46, 49, 74

composition, proportional, 95, 97

cosmological time, 6, 45, 50–51, 61–62, 71, 73, 80, 181–82, 185–86; absolute, 6, 46–47, 62, 181, 185; arrow of, 60–61; unidirectional, 45, 62

cosmology, 5, 14

cues, environmental, 49, 179

cyclical time, 60, 62

dispersion: degree of, 146, 149, 177; temporal, 151, 163–64

double brooding, 105–6; rate of, 105–6

duration: decreased, 56, 58; increased, 53, 56

duration of critical phenophases, 47, 111

duration of flowering, 51, 163, 165

duration of phenophases, 55, 57, 70, 113–14, 169

ecological time, 43, 59–62, 65, 80, 104–6, 121, 167, 169–70, 176; availability of, 99, 104, 126, 167, 169, 171, 176, 178; availability of relative, 45, 136–38, 157–58, 181–82; increasing availability of relative, 155, 157

ecology, metabolic theory of, 46, 198

effects, latency, 68, 70

emergence: absolute timing of, 101; annual timing of, 101, 103, 113, 128, 184; community-level median, 119–20; community-wide, 154; earlier onset of, 100, 129; earlier species-level, 132; mean timing of, 102–3, 130; median timing of, 118; order of, 99, 101, 103; rank order of, 100–103, 174; species' rank order of, 101; state of, 96, 113; timing of, 71–72, 96, 99, 101–3, 113, 115,

emergence (*continued*)
   117–19, 121, 123, 125–32, 141, 144, 147,
   153–54, 183–84
emergence dates, 51, 71–72, 101, 103, 115,
   117, 119–22, 126, 128–32; annual, 153;
   dispersion of, 118, 123, 129, 134, 137
emergence phenology, 71–72, 103, 117, 121,
   138, 143, 165
emergence range, 129
energy, 4, 14, 46, 168, 182
ENSO (El Niño Southern Oscillation), 17, 162,
   174
environmental conditions: expansion of realized,
   88, 92; function of changes in realized, 87, 91
environmental constraints, 42, 60, 68, 114–16,
   118, 121, 123–25, 127, 133; alleviation of, 2,
   53, 135, 140
environments, two-dimensional, 109, 111
estimates of rates of change, 29–30
estimates of rates of phenological change, 27, 30
event cones, 13
events, timing of, 2, 6, 10, 25, 44, 55
evolution of life history strategies, 4, 6, 16–17
extinction, 45, 64, 88

favorable conditions: advanced timing of, 85,
   87; delayed timing of, 85, 87
flowering, 1, 5, 20, 25, 35–36, 51, 54, 65,
   106–7, 113, 123–25, 162–65, 169–71, 174,
   176–78; annual periods of, 70, 177; annual
   timing of, 20, 164; community-wide, 166,
   176; earlier, 164–65; seasonal, 66, 174; timing of, 50–51, 63, 123, 138–39, 163–64, 171,
   174, 178, 184, 219
flowering dates, first, 124, 159
flowering events, multiple, 65–66
flowering phenology, 84, 123, 138, 162–65,
   171, 176
forbs, 16, 95–96, 154
fragmentation of time, 64–65
fruiting phenology, 123, 173, 175
fundamental phenological niche, 80–82, 85,
   87–89, 91–92

generalist associations, 89, 93
generalist pollinator-plant mutualisms, 84, 91
graminoids, 95–96, 154
growing season length, 176

habitat fragmentation, 64–65
hatching, 35, 54, 162
herbivore arrival, 155, 158

idealized life history cycles, 54, 65
insect pollinators, wild, 83
interactions: competitive, 62, 106; consumer-
   resource, 62, 74, 106, 112, 134, 178; exploitation, 74, 112, 134, 178; mutualistic, 49,
   134; organism-organism, 47, 74
interspecific competition, 54, 112, 117, 119, 163
interspecific competition in time, magnitude
   of, 118
interspecific dispersion, 131, 150
intraspecific aggregation, 126
intraspecific competition, 108–9, 114, 117, 119,
   158; magnitude of, 118
intraspecific dispersion of dates, 149
intraspecific phenological dispersion, 149, 151,
   153, 155, 157, 177
IPCC, 17, 20, 24, 30, 176; Fifth Assessment
   Report, 20, 24

Kangerlussuaq study site, 95, 97, 99–100, 103,
   120, 123–24, 143–44, 153

land-surface warming, rate of, 41
late-season life history strategists, 125
leaf emergence, 96–97
life history cycles, 46, 51, 53–56, 64, 96,
   112–13; annual, 15, 55, 59
life history dynamics, 1, 127, 169, 179
life history events: timing of, 51, 53, 55–57,
   108, 111–12, 127; timing and duration of, 53;
   timing of expression of, 42, 53, 108–9, 111
life history stages: critical, 58, 112; successive,
   65, 112
life history strategies: early-emergence, 121;
   early-onset, 103; individual, 67, 97, 103,
   117; late-season, 70, 123; respective, 16, 25;
   vernal, 16, 98
life history timing, 182, 184
life history traits: expression of, 2, 44, 48,
   51–52, 62, 67, 107, 117, 124, 131, 178, 180;
   timing of, 44, 183; timing of expression of,
   35, 62, 64, 67, 71, 103, 106, 108–10, 118,
   121, 125, 127, 133, 135, 182–85
light availability, 170, 175
local extinction, 85, 87–88, 91–94, 104, 139

metabolic scaling laws, 3, 50
migration, 21, 35–36, 50, 54, 62, 64, 144, 181;
   long-distance, 48, 181
migration timing, 50, 62, 144
migratory birds, 15, 50, 54, 56
mismatch, 106, 138, 163–64, 182

moisture availability, 170–71, 174, 179
moisture limitation, 111, 174–75
muskoxen, 143–45, 146, 149–51, 153–55, 157–61
mutualisms, 49, 62, 82, 84–85, 87–88, 91–93, 112, 138
myrtaceous trees, 176–77

natural selection, 47, 108
nature of time, 7, 9, 42
nesting phenology, 105–6
niche, fundamental, 81, 87
niche breadth, 69; relative phenological, 75, 77, 79
niche complementarity, 47, 91–92
niche packing, 104
niche space: fundamental relative phenological, 79–80; temporal, 68, 70
null infinity, 13

observational studies, 124–25, 163
onset of activity, timing of, 98–99

partitioning space, 63–64
past, relative, 61–62
pathogens, 83, 112
phenological advance: differential rates of, 127, 162; greatest rates of, 20, 22, 26, 165; rates of, 21–22, 25–26, 29–31, 33–34, 37, 39–40, 42, 98, 107, 123, 130, 161
phenological aggregation, 178
phenological cascades, 142, 153, 155, 158, 162; direct, 142, 155, 157; indirect, 142, 166
phenological community, 8, 12, 67, 94–101, 103–4, 106, 112, 115, 118–19, 125–26, 166, 170, 185
phenological community concept, 95–96
phenological delays, 23, 25, 29, 33, 58, 132, 161, 180
phenological dispersion, 132–33, 135, 137, 140–41, 144, 146, 157, 161, 176–77; degree of, 146, 149; increased, 171, 176
phenological dynamics, 15–17, 26, 31, 33, 44–45, 47, 107, 126–27, 133, 139–40, 142–43, 162–66, 169–71, 174–75, 180–81; delayed, 23
phenological mismatch, 84, 138, 164–65, 181–82; consequences of, 139, 165
phenological mutualism, 88, 91
phenological niche, 66–68, 70, 79–80, 82, 85, 89, 91–92; realized, 81

phenological niche axes, 79, 104
phenological niche breadth, 70, 83
phenological niche concept, relative, 71, 76
phenological niche conservatism, 68, 70, 83, 85; consequences of, 92–93
phenological niche invasion, 68, 70
phenological niche space, 82, 88–89
phenological overlap: degree of, 158, 161; extent of, 159
phenological plasticity, 71, 103, 159, 163
phenological segregation, 106, 176–78
phenological stasis, 6, 23, 25–26, 48, 58, 139, 141, 180
phenological trends, 20, 22, 26–29, 33, 35–36, 38; estimates of, 26–29, 31, 33; positive, 22–23
phenological variation, 170, 175
phenology, study of, 1, 48–49, 167, 187
phenophases, 5–6, 45–46, 54–55, 57, 70, 94–97, 113–14, 119, 123–27, 169, 171, 174, 184, 186
photoperiod, 2, 47–48, 104, 167, 178–79, 181–85
pollination, 35–36, 65, 84
pollinator community, 165
pollinator diversity, 84, 164
pollinator persistence, 89
pollinator-plant associations, 89, 163
pollinators, 51, 66, 84–85, 87–89, 91–93, 138–39, 162–66, 177–78, 189; multiple potential, 87, 91; specialist, 84–85, 87
pollinator services, 164, 174, 177
population dynamics, 9, 38, 51–52, 143, 180
priority effects, 68–70
Pyrola grandiflora, 100, 131

rank order series, 99–100
realized environmental conditions, 81–82, 85, 87–89, 91–93
recurrent time, 6, 50–51, 60–62, 181–82, 185
relationism, 7–8
relative ecological time, 6, 8–9, 12–13, 45, 47, 50–51, 60, 62–63, 66–67, 106, 136–39, 157–58, 178–79, 181–82, 185–86
relative phenological niche, 71, 73–75, 77–82
relative timing, 51, 54, 61, 98
resource availability, 15, 17, 53, 105, 153, 155
resource competition, 74–75, 80
resource phenology, 66, 75, 79, 82, 105, 133–34, 153, 159, 162
resource species, 54, 76, 78, 133–34, 138, 169, 178

Salix glauca, 100–101, 131
seasonal cycles, 17, 48
seasonal dynamics, 75, 77
seasonal growth, 48, 135
seasonality, 15–18, 35, 51, 60, 167, 179–80, 182
seasonal onset, 73, 75
seasonal timing, 63, 73, 75, 77, 139, 163, 170, 177, 182; optimal, 80, 82
season of biological activity, 71, 74
segregation, increasing, 126–27
shrubs, dwarf, 95, 101
snow melt timing, 3, 185
solar insolation, 16, 167, 179
solar irradiance, 2–3, 16–17, 47–48, 73, 167, 171, 174, 179, 182–83, 185
space: role of, 48, 59; scales of, 44, 51–52; use of, 44, 59–60, 63, 181; utilization of, 63–64
space-time, 11–13, 14, 52, 59, 63
specialist pollinator-plant mutualisms, 87–88
species: congeneric, 43, 177; mutualistic, 112, 134; resident, 98, 143
species interactions, 55, 67, 82, 97, 112, 181, 185; horizontal, 107, 112
spring arrival, 24, 54
spring emergence, annual timing of, 20, 35

spring migration, 20, 53–54
springs: cold, 105–6; false, 65, 195

tallgrass prairie, 123–24
tawny owls, 162
temperature, annual, 16, 175
temperature anomalies, 38–39
temperature limitation, 35, 65, 108–9, 184
temporal dependence, 49, 151
temporal segregation, increased, 131–32
theory of time: dynamic, 10, 62; static, 11
timing and duration, 46, 53, 75, 80, 107, 161
tree swallows, 105
tritrophic-level systems, 139, 141–42, 159
tropical species, 170, 174–75, 178
tropical systems, 17, 96, 106, 132, 163, 166–67, 169–71, 175–78

unidirectionality of time, 42–43
use of relative ecological time, 178–79
use of time, 5, 46–47, 49, 58, 60, 64, 66, 73, 111–12, 131, 133–34, 169, 177–78, 182, 184–86

variation, latitudinal, 17, 37–38, 167, 172–73
vertical species interactions, 112, 133–34, 138

MONOGRAPHS IN POPULATION BIOLOGY

EDITED BY SIMON A. LEVIN AND HENRY S. HORN

1. *The Theory of Island Biogeography* by Robert H. MacArthur and Edward O. Wilson
2. *Evolution in Changing Environments: Some Theoretical Explorations* by Richard Levins
3. *Adaptive Geometry of Trees* by Henry S. Horn
4. *Theoretical Aspects of Population Genetics* by Motoo Kimura and Tomoko Ohta
5. *Populations in a Seasonal Environment* by Steven D. Fretwell
6. *Stability and Complexity in Model Ecosystems* by Robert M. May
7. *Competition and the Structure of Bird Communities* by Martin L. Cody
8. *Sex and Evolution* by George C. Williams
9. *Group Selection in Predator-Prey Communities* by Michael E. Gilpin
10. *Geographic Variation, Speciation, and Clines* by John A. Endler
11. *Food Webs and Niche Space* by Joel E. Cohen
12. *Caste and Ecology in the Social Insects* by George F. Oster and Edward O. Wilson
13. *The Dynamics of Arthropod Predator-Prey Systems* by Michael P. Hassel
14. *Some Adaptations of Marsh-Nesting Blackbirds* by Gordon H. Orians
15. *Evolutionary Biology of Parasites* by Peter W. Price
16. *Cultural Transmission and Evolution: A Quantitative Approach* by L. L. Cavalli-Sforza and M. W. Feldman
17. *Resource Competition and Community Structure* by David Tilman
18. *The Theory of Sex Allocation* by Eric L. Charnov
19. *Mate Choice in Plants: Tactics, Mechanisms, and Consequences* by Nancy Burley and Mary F. Wilson
20. *The Florida Scrub Jay: Demography of a Cooperative-Breeding Bird* by Glen E. Woolfenden and John W. Fitzpatrick
21. *Natural Selection in the Wild* by John A. Endler
22. *Theoretical Studies on Sex Ratio Evolution* by Samuel Karlin and Sabin Lessard
23. *A Hierarchical Concept of Ecosystems* by R. V. O'Neill, D. L. DeAngelis, J. B. Waide, and T.F.H. Allen
24. *Population Ecology of the Cooperatively Breeding Acorn Woodpecker* by Walter D. Koenig and Ronald L. Mumme
25. *Population Ecology of Individuals* by Adam Lomnicki
26. *Plant Strategies and the Dynamics and Structure of Plant Communities* by David Tilman

27. *Population Harvesting: Demographic Models of Fish, Forest, and Animal Resources* by Wayne M. Getz and Robert G. Haight

28. *The Ecological Detective: Confronting Models with Data* by Ray Hilborn and Marc Mangel

29. *Evolutionary Ecology across Three Trophic Levels: Goldenrods, Gallmakers, and Natural Enemies* by Warren G. Abrahamson and Arthur E. Weis

30. *Spatial Ecology: The Role of Space in Population Dynamics and Interspecific Interactions*, edited by David Tilman and Peter Kareiva

31. *Stability in Model Populations* by Laurence D. Mueller and Amitabh Joshi

32. *The Unified Neutral Theory of Biodiversity and Biogeography* by Stephen P. Hubbell

33. *The Functional Consequences of Biodiversity: Empirical Progress and Theoretical Extensions*, edited by Ann P. Kinzig, Stephen J. Pacala, and David Tilman

34. *Communities and Ecosystems: Linking the Aboveground and Belowground Components* by David Wardle

35. *Complex Population Dynamics: A Theoretical/Empirical Synthesis* by Peter Turchin

36. *Consumer-Resource Dynamics* by William W. Murdoch, Cheryl J. Briggs, and Roger M. Nisbet

37. *Niche Construction: The Neglected Process in Evolution* by F. John Odling-Smee, Kevin N. Laland, and Marcus W. Feldman

38. *Geographical Genetics* by Bryan K. Epperson

39. *Consanguinity, Inbreeding, and Genetic Drift in Italy* by Luigi Luca Cavalli-Sforza, Antonio Moroni, and Gianna Zei

40. *Genetic Structure and Selection in Subdivided Populations* by François Rousset

41. *Fitness Landscapes and the Origin of Species* by Sergey Gavrilets

42. *Self-Organization in Complex Ecosystems* by Ricard V. Solé and Jordi Bascompte

43. *Mechanistic Home Range Analysis* by Paul R. Moorcroft and Mark A. Lewis

44. *Sex Allocation* by Stuart West

45. *Scale, Heterogeneity, and the Structure of Diversity of Ecological Communities* by Mark E. Ritchie

46. *From Populations to Ecosystems: Theoretical Foundations for a New Ecological Synthesis* by Michel Loreau

47. *Resolving Ecosystem Complexity* by Oswald J. Schmitz

48. *Adaptive Diversification* by Michael Doebeli

49. *Ecological Niches and Geographic Distributions* by A. Townsend Peterson, Jorge Soberón, Richard G. Pearson, Robert P. Anderson, Enrique Martínez-Meyer, Miguel Nakamura, and Miguel Bastos Araújo

50. *Food Webs* by Kevin S. McCann

51. *Population and Community Ecology of Ontogenetic Development* by André M. De Roos and Lennart Persson

52. *Ecology of Climate Change: The Importance of Biotic Interactions* by Eric Post

53. *Mutualistic Networks* by Jordi Bascompte and Pedro Jordano

54. *The Population Biology of Tuberculosis* by Christopher Dye

55. *Quantitative Viral Ecology: Dynamics of Viruses and Their Microbial Hosts* by Joshua Weitz

56. *The Phytochemical Landscape: Linking Trophic Interactions and Nutrient Dynamics* by Mark D. Hunter

57. *The Theory of Ecological Communities* by Mark Vellend

58. *Evolutionary Community Ecology: The Dynamics of Natural Selection and Community Structure* by Mark A. McPeek

59. *Metacommunity Ecology* by Mathew A. Leibold and Jonathan M. Chase

60. *A Theory of Global Biodiversity* by Boris Worm and Derek P. Tittensor

61. *Time in Ecology: A Theoretical Framework* by Eric Post